THE SPECIAL FORCES

Other National Historical Society Publications:

THE IMAGE OF WAR: 1861-1865

TOUCHED BY FIRE: A PHOTOGRAPHIC PORTRAIT OF THE CIVIL WAR

WAR OF THE REBELLION: OFFICIAL RECORDS
OF THE UNION AND CONFEDERATE ARMIES

OFFICIAL RECORDS OF THE UNION AND CONFEDERATE NAVIES
IN THE WAR OF THE REBELLION

HISTORICAL TIMES ILLUSTRATED ENCYCLOPEDIA OF THE CIVIL WAR

CONFEDERATE VETERAN

THE WEST POINT MILITARY HISTORY SERIES

IMPACT: THE ARMY AIR FORCES' CONFIDENTIAL HISTORY
OF WORLD WAR II

HISTORY OF UNITED STATES NAVAL OPERATIONS IN WORLD WAR II
by Samuel Eliot Morison

HISTORY OF THE ARMED FORCES IN WORLD WAR II
by Janusz Piekalkiewicz

A TRAVELLER'S GUIDE TO GREAT BRITAIN SERIES

MAKING OF BRITAIN SERIES

THE ARCHITECTURAL TREASURES OF EARLY AMERICA

For information about National Historical Society Publications, write:

The National Historical Society, 2245 Kohn Road, Box 8200,
Harrisburg, Pa 17105

THE ELITE
The World's Crack Fighting Men

THE SPECIAL FORCES

Ashley Brown, Editor

Jonathan Reed, Editor

A Publication of
THE NATIONAL HISTORICAL SOCIETY

Published in Great Britain in 1986 by Orbis Publishing

Special contents of this edition copyright © 1989 by the
National Historical Society

Library of Congress Cataloging-in-Publication Data
The Special forces / Ashley Brown, editor , Jonathan Reed, editor.
    p.   cm.—(The Elite : the world's crack fighting men ; v. 13)
    ISBN 0-918678-51-X
    1. Special forces (Military science)—History—20th century.
2. Special operations (Military science)—History—20th century.
I. Brown, Ashley. II. Reed, Jonathan. III. National Historical
Society. IV. Series: Elite (Harrisburg, Pa.) ; v. 13.
U262.S64   1990
355′.00904—dc20   89-12204
                                                        CIP

CONTENTS

INTRODUCTION

Resisting massive amphibious invasions. Combatting terrorists. Countering rebels by horse-mounted infantry. All have been the province of some very unusual and daring men who were outstanding enough to be accepted for membership in the SPECIAL FORCES. Often small in number, pitted against heavy odds, and fighting on with courage alone, they have earned their place among the ELITE.

Consider the 22d Bataillon de Chasseurs Alpins. Literally mountain climbers, these denizens of the French Alps carried weapons into the craggy recesses as well. But in 1956, they took their climbing gear and their rifle to the Great Kabylia in Algeria, facing rebels in strange territory, under a blazing sun never experienced in the Alps. Mountaineering was the special skill of other units, as well. In World War II the United States Army sent the 10th Mountain Division to scale the Riva Ridge in Northern Italy. In 1940, when Hitler invaded Norway, he looked to his 3d Mountain Division, composed of hardy Austrians.

Others who didn't climb mountains, climbed into saddles instead, like Grey's Scouts, a band of picked Rhodesians who became mounted infantry in the struggle for the bush in the 1970s. Around the world, beneath the jungle covering, other brave bands of men lived and fought in tunnels, the Viet Cong guerrillas. Then there were the incomparable Grenadier Guards. In Tunisia, in 1943, the 6th Battalion met the Afrika Korps in a brutal ring of hills called The Horseshoe.

On every inhabited continent in the world, the SPECIAL FORCES have born the brunt of some of our fieriest trials, especially the peacetime ''combat'' against terrorists. Called by a host of names, from Delta Force to GEO, they have been the protectors of thousands in an increasingly hostile world.

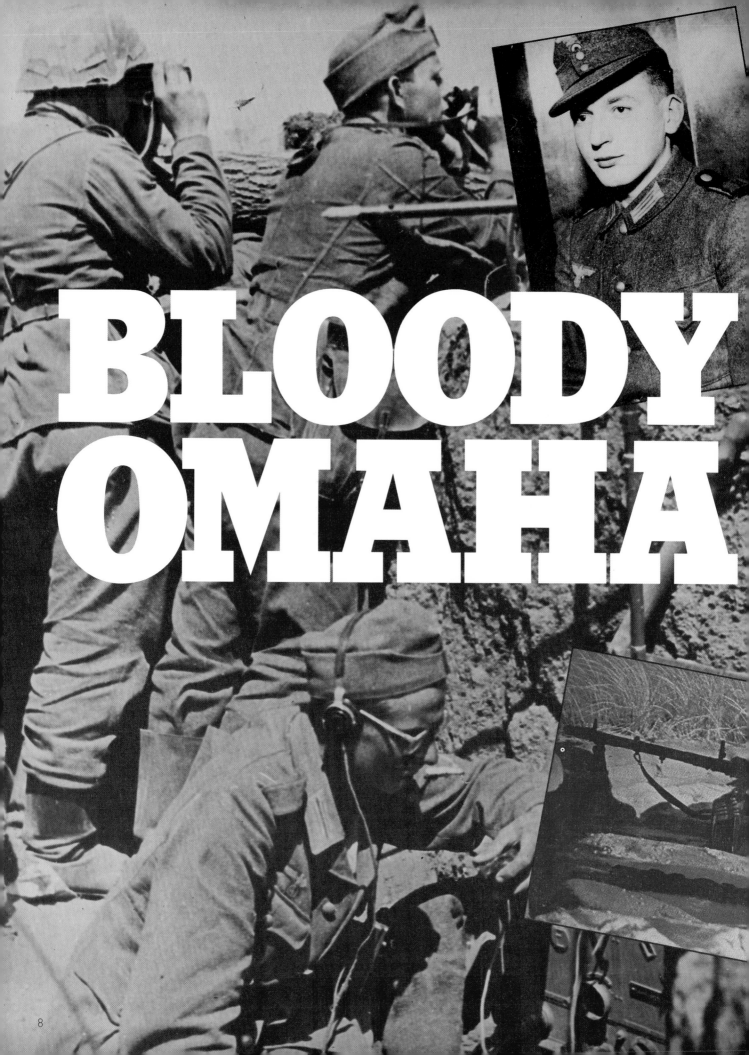

BLOODY OMAHA

8

When, at dawn on 6 June 1944, waves of American infantrymen began to pour onto Omaha Beach in Normandy, they were stopped short by the 352nd Grenadier Division

IN THE QUIET DAWN of 6 June 1944, Gefreiter (Lance-Corporal) Heinz Severloh, a 21-year old 'farmer in uniform', stood up and looked out over the grey waters of the English Channel. His post was Widerstandnest (pocket of resistance) 62, a bunker dominating the beach at Colleville-sur-Mer, and before him lay the familiar outline of his MG-42 machine gun, a formidable weapon capable of spewing out 1000 rounds a minute. One of hundreds of such positions, the bunker was part of the westward bastion of Hitler's 'Fortress Europe'. Strewn with obstacles, the stretch of sand before it was destined to go down in history as the killing ground of over 3000 American soldiers. Its name was Omaha Beach.

The defences in Normandy had been begun in earnest early in April 1944. The German tactical plan against amphibious attack was to hold the Allied forces at the water's edge until mobile reserves arrived to finish them off. The troops were to be held by obstacles and mines on the tidal flat and the beach shelf, combined with concentrated crossfire.

The first band of static obstructions consisted of a series known to the Allies as 'Element C'. These were gate-like reinforced steel frames with iron supports on rollers, positioned about 250yds out from the high-water line. The main support girders were about 10ft high and waterproofed Tellermines were lashed to the uprights. Between 20 and 25yds behind these was a second band of obstacles, this time heavy logs driven into the sand at an angle, with the mine-tipped ends pointing seaward. There were

Left: Artillery observation posts were a vital element of the Atlantic Wall defences on Omaha Beach. Concrete pillboxes containing artillery, anti-tank weapons and machine guns were supplemented by simple machine-gun positions such as the MG-34 nest (below). Above left: Gefreiter Heinz Severloh's unit, the 352nd Grenadier Division, comprised both raw recruits and experienced veterans (right).

also log ramps, reinforced and mined. The final row consisted of 'hedgehogs', each a structure of three or more steel rails or angles, crossed at the centre. Nearly six feet high, they were set so firmly that the ends would be able to stave in the bows of landing craft.

Beyond the shingle were the mines. Some of these sprang up on contact and exploded at waist height, blasting out hundreds of steel balls. Concertina wire was laid to snag and hold the attackers, and fougasses (crude 'mortars' consisting of holes in the rock filled with stones and lumps of iron laid over explosive charges) were laid and attached to trip wires and to the concertina wire.

The tidal flat and beach shelf were well covered by firing positions. On Omaha Beach, observation and flanking fire from cliff positions at either end were aided by the curving crescent of the shoreline. Lethal enfilading fire could also be set up from

352ND GRENADIER DIVISION

The 352nd Grenadier Division was raised on 22 September 1943, based in Wehrkreis (Army Corps Area) XI and with its headquarters in Hanover. It was brought to combat readiness in February and March 1944 as a Neue Aufstellung (newly formed) division and consisted of the 914th, 915th and 916th Grenadier Regiments, 352nd Artillery Regiment and 352nd Signals Unit. At the time of the Normandy invasion, the division was one of the 12 in the Seventh Army, commanded by Generaloberst (General) Friedrich Dollmann, and formed part of Army Group B under Field Marshal Erwin Rommel. He in turn was under the command of Field Marshal Gerd von Rundstedt, the Commander-in-Chief West. Following Rommel's injury and death in 1944, Army Group B eventually came under the command of Field Marshal Walther Model. The 352nd Division was on the left flank in Normandy on 6 June 1944, defending Omaha Beach. Major Werner Pluskat of 352nd Artillery Regiment was one of the very first to see the invasion fleet, in his case Force O (for Omaha). Although the 352nd Division ceased to exist in August 1944, other elements of the Seventh Army managed to escape from the Falaise pocket and, together with the Fifth Panzer Army, to take part in the Battle of the Ardennes in December 1944. Although the 352nd Volks-Grenadier Division was included in the Ardennes forces, this unit was the renumbered 581st Volks-Grenadier Division and was in no way connected with the formation in Normandy. Above: The silver close-combat badge of the Wehrmacht.

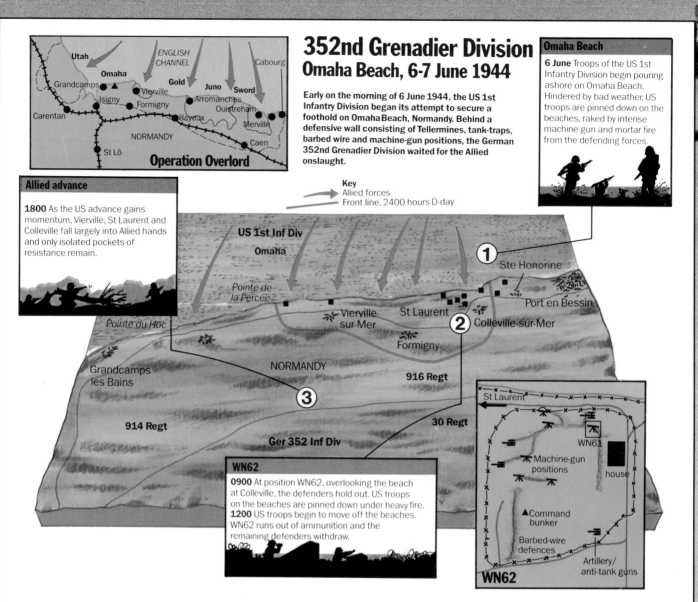

352nd Grenadier Division
Omaha Beach, 6-7 June 1944

Early on the morning of 6 June 1944, the US 1st Infantry Division began its attempt to secure a foothold on Omaha Beach, Normandy. Behind a defensive wall consisting of Tellermines, tank-traps, barbed wire and machine-gun positions, the German 352nd Grenadier Division waited for the Allied onslaught.

Omaha Beach

6 June Troops of the US 1st Infantry Division begin pouring ashore on Omaha Beach. Hindered by bad weather, US troops are pinned down on the beaches, raked by intense machine-gun and mortar fire from the defending forces.

Key
Allied forces
Front line, 2400 hours D-day

Allied advance

1800 As the US advance gains momentum, Vierville, St Laurent and Colleville fall largely into Allied hands and only isolated pockets of resistance remain.

US 1st Inf Div
Omaha

Pointe de la Percée

Ste Honorine

Port en Bessin

Vierville sur-Mer

St Laurent

Colleville-sur-Mer

Formigny

Pointe du Hoc

NORMANDY

916 Regt

Grandcamps les Bains

914 Regt

Ger 352 Inf Div

30 Regt

WN62

0900 At position WN62, overlooking the beach at Colleville, the defenders hold out. US troops on the beaches are pinned down under heavy fire. **1200** US troops begin to move off the beaches. WN62 runs out of ammunition and the remaining defenders withdraw.

St Laurent

WN61

Machine-gun positions

house

▲Command bunker

Barbed-wire defences

Artillery/ anti-tank guns

WN62

emplacements between Vierville and Pointe de la Percée. Taken together with numerous artillery emplacements to the rear, all protected by a well-developed system of intercommunicating bunkers, Gefreiter Severloh and his fellow soldiers of the 916th Grenadier Regiment, 352nd Grenadier Division, could count on their positions being taken only after an extremely costly battle to their enemy.

Severloh's division, like many of those manning the Atlantic Wall, was newly formed from a mixture of new recruits and veterans of the Eastern Front, most

Top right: US infantry hug the ground between obstacles on 'Easy Red' beach at Omaha. Above right: Members of the 2nd Battalion, 916th Grenadier Regiment, before the battle. Below: The crew of a 2cm flak half-track waits in a copse for an American breakthrough.

of them survivors of the 389th Grenadier Division, which had been virtually destroyed as a fighting unit by the Red Army in the Stalingrad offensives. The survivors had been transferred to Normandy to be re-formed into what became the 352nd Grenadier Division, their old 546th Grenadier Regiment providing staff for the 1st Battalion, 916th Grenadier Regiment, as well as a pool of troops for the 1st and 2nd Battalions of that new regiment, which was officially formed on 14 November 1943.

The division was billeted at St Lô and had spent several months there working up to full strength. This had given the veterans, who were glad to be given a break from the horror of the Eastern Front, a chance to impart some of their painfully learned combat skills to the new men. By the time the 352nd Grenadier Division was moved up in May 1944 to the strip of coastline between Grandcamps and Arromanches, earmarked for them by Field Marshal Rommel, they had become a cohesive fighting force of well above-average capability.

The main problems faced by the 352nd were shortages of munitions, motorised transport and fuel. As soon as the division's units had taken up their positions they had begun to improve and increase the beach defences. Spurred on by Rommel, they

had cut stakes from the Cerisy Forest, 11 miles inland, carted them to the beach and then driven them in by hand. Despite their difficulties, by 6 June the defences were excellent, and it was these that promised to turn a 'walkover' into a bitterly disputed battle.

Gefreiter Heinz Severloh had joined the division in November 1943 after serving with the 321st Grenadier Division on the Eastern Front. At 0130 hours on 6 June, following an alarm from Major Werner Pluskat, his platoon had been rushed from its comfortable quarters at a farmhouse in nearby Houteville. The men in his bunker had already been alerted by Feldwebel (Senior Sergeant) Pieh. Severloh insists that he was the first to see the ships:

'Five, maybe seven ships running parallel to the coast laying a smokescreen. They were only visible for one minute, but visible to the naked eye. I called Oberleutnant [First Lieutenant] Freking, my commander, who also saw them. All the reports I know claim we never saw these ships.'

Freking told Severloh: 'Get the machine gun ready. Hit them when they're in the water'

Freking ordered signallers to attempt to contact the ships. There was no answer, and the troops became certain that they were enemy vessels. Freking then picked up his binoculars, looked out and said, 'Oh man, they're here!' Severloh confirmed the fact: 'We all saw it. It was a city, no water, just a great, massive, metal-grey city with those strange aerial balloons heading steadily towards us.'

In a frighteningly short space of time the American landing craft were beginning to hit the beach. Severloh remembered that awful moment:

'They were coming right towards our guns. It was ebb tide, which meant they had to wade helplessly through shoulder-deep water. The first landing boat started towards us. We were standing on top of the bunker and they could have seen us clearly if they had had time to look.'

Two hundred yards from the shoreline, the enemy were leaping from landing craft and wading ashore. The commander of the 916th Grenadier Regiment, Oberst (Colonel) Goth, ordered the men to hold their fire until the enemy was at the water's edge. The grenadiers waited. Freking told Severloh: 'Get the machine gun ready. Hit them when they're in the water, before they have time to spread out. Not a minute earlier.' The Americans had their rifles on their backs and couldn't defend themselves.

'Fire!' yelled Feldwebel Krone, who was sitting close by in the entrance to the observation bunker of the 1st Battalion, 352nd Artillery Regiment. Severloh opened up with his ferocious weapon and watched with a kind of dreadful fascination as his bursts began to cut down the first assault wave. By mid-morning he had fired over 12,000 rounds. Over 240 50-round belts had been expended. Severloh recalled:

'I fired the usual pattern, going from left to right and then up the ramp as they came down. Then we stopped firing and I was handed a rifle and shot at those swimming in the water. It was slaughter. I don't know how many were on the boat, but I doubt if 100 of them made it to land.

'At one moment a jeep with a .50 cal. machine gun raced up the beach. Leutnant [Lieutenant] Grass and Feldwebel Pieh moved forward to the edge of the trench with a rifle and grenade launcher – the first shot blew the jeep to pieces and the next shot disabled a Sherman tank.'

At 0900 hours, Oberst Goth sent a report to General Marcks at LXXXIV Corps HQ at St Lô:

'All along the front the enemy are looking for cover and hiding behind obstacles...many vehicles are burning on the shore, also 10 tanks. Their engineers are unable to work. All disembarkation has ceased. The ships offshore have ceased firing. The fire of our artillery and guns is on target and inflicting heavy casualties on the enemy, a large number of whom lie dead on the beaches.'

At 1030, the 1st Battalion, 914th Grenadiers, was having a hard time containing troops from the US 116th Regiment and Company B, 5th Ranger Battalion, who were fighting their way into Vierville after clearing a beach exit. Three truckloads from the 914th also caught about 25 Americans who barricaded themselves into the château of Le Vaumicel. After some heavy exchanges of fire the Germans withdrew and, although reinforced, they were unable to hold onto Vierville and were gradually pushed out to a few houses on the southern outskirts.

WHERE WILL THEY LAND?

In December 1943, Hitler told his generals that, 'If they attack in the west, that attack will decide the war.' The great question was where the landing would take place, and although German intelligence knew that attack was imminent in June 1944, the location of the landing zone area remained a well-kept secret. As it was, in June Hitler had only 59 divisions in France and the Low Countries, compared with 165 deployed on the Eastern Front, and these had to defend 1000 miles of coastline, from the northern tip of the Netherlands to the mouth of the Loire.

The Allies used a great many devious means to convince German intelligence that the landing would take place in the Pas-de-Calais, and in this they were very successful. Hitler was convinced that the Allies would land there and, although Field Marshals Rommel and von Rundstedt suspected that Normandy would be the site, no more than minimal reinforcements were ordered into the area.

Hitler's position in Normandy would have been much stronger had he not been convinced that any attack in Normandy would be a feint. Rommel had believed it vital to place his panzer divisions on the coast, but had not been permitted to do so. Thus, only the weak 21st Panzer Division, based south of Caen, was within reach of the beaches. Had Rommel's request to place a second panzer division near St Lô been granted, the division could have had a decisive influence on the outcome of the landings. Instead, the burden of defeating the invasion fell mainly on the soldiers manning the static defences of the Atlantic Wall and the close reinforcements stationed in nearby coastal towns and villages.

Engineer Brigade Group which, acting as surrogate tanks, pushed their way towards St Laurent.

The 2nd Battalion, 726th Grenadier Regiment, which was stationed alongside the 352nd Division's units, began to fall back in disorder, having sustained heavy casualties and having had most of its blockhouses reduced to rubble by naval destroyers which had moved inshore to engage the bunkers at very close range. Meanwhile, the 2nd Battalion, 916th Grenadiers stood firm, repulsing all attacks and mopping up small groups of Americans who had found their way through the dunes by using areas

Feldwebel Mayer, 916th Regiment, 352nd Grenadier Division, Normandy 1944

Senior Sergeant Mayer is wearing a reed green denim combat jacket over field grey woollen trousers. His steel helmet is fitted with a crude wire mesh for the insertion of foliage as camouflage. On his back is the German cylindrical gas mask container (by now more likely to contain food than a gas mask), with a poncho in camouflage material, a green canvas bread bag and a water bottle with cup. His weapon is the German 7.92mm Kar 98k rifle.

Elsewhere, elements of the 916th were engaged in house-to-house fighting with the US 115th and 116th Infantry Regiments in Colleville and St Laurent, who had infiltrated off the beach through the minefields between the exits of Les Moulins and Le Ruquet. Around midday, the bunkers of Le Ruquet were knocked out by combined fire from a solitary tank of the 18th Regimental Combat Team (RCT) and a destroyer offshore. Up this gulley flooded long files of troops from the US 115th Infantry Regiment, crouching low behind the bulldozers of the Special

where the mines had been detonated by the preliminary offshore bombardment.

But however hard the men in the individual units of the 352nd Grenadier Division fought to prevent the Americans from gaining a foothold, they had no prospect of holding back the huge wave of men and armour bearing down on them. Shrapnel hit Severloh's machine gun, tearing off the sights and hurtling fragments into his face. By this time his ammunition was all but exhausted anyway. Freking sent a last message to his men: 'Gunfire barrage on the beach – we're pulling back.' The defenders all shook hands. Freking said: 'Heinz, you go first, I'll be behind you.' The men withdrew under a hail of fire, jumping from one crater to another. Freking, Grass and Pieh never made it. Momentarily in the open, they were cut down by American fire.

Two years after the end of the war he wrote to his old landlord at his French billet, asking where he could find Freking's battlefield grave. It was located and Freking's body was moved to the German military cemetery of La Cambe.

Surviving his group's move to the rear, Severloh teamed up with a radio operator, but they then drew fire and were both wounded:

'The bullets passed through both the radioman's buttocks and went through my pay records book,

by 1800 hours, with heavy casualties being inflicted by the US 18th and 16th Infantry Regiments. The German units in Colleville were surrounded by 2000 hours, and late in the evening the road from Colleville to St Laurent was cut. Some of the blockhouses defending the main gullies leading off Omaha Beach were still in action, however, and they kept up a continual harassing fire on the waves of troops following up from the sea. In the end, slowly but surely, the troops of the 916th Grenadiers were pushed back to the southern edge of the high bluffs overlooking Colleville and Le Grand Hameau.

By midnight, the Americans had still made relatively little territorial gain, but their sheer numbers were beginning to overwhelm the battle-weary defenders. The line of defence was stretched but not broken; cracks had appeared, and without reinforcement the defence was bound to crumble. But in a day of combat against vastly superior forces, the 352nd Grenadier Division had fought extremely well.

Not long ago, Heinz Severloh was invited to inspect an MG-42 on show in a museum of the war. He declined, saying it brought back sad memories. He was genuinely moved when watching the original documentary footage of Omaha Beach, hearing the announcer describe the withering machine-gun fire

Top: Artillery officers of the 352nd Grenadier Division. Carrying a riding crop in the foreground is Major Werner Pluskat, one of the first to warn of the invasion fleet. Above left: Two young recruits of the division, Wilhelm Stetter (above) and Albert Summerer (below). Left: Following the capture of a key bunker, Staff Sergeant Jack Scarborough looks down on the corpse of one soldier who died for his Fatherland. Above: Against the vast resources of the Allies, the Germans on the Normandy beaches had little hope of success. Here, a casualty is carried into captivity.

hit me in the hip and knocked me three metres away. We then found some of our division's soldiers, halfway between the beach and Colleville.' Given some medication, he remembers not feeling any pain. He was given a rifle and put in charge of some American prisoners.

Speaking to him in the German tongue of his parents, one of the Americans asked Severloh when they were going to be shot. Severloh reassured him, telling him that no-one was going to be shot, for such an action would bring dishonour to his unit. Rounding up the prisoners and several wounded, he attempted to join up with German forces to the rear of the beachhead. Coming under fire and realising that it would be impossible to make progress, he told the Americans that he wished to surrender.

By mid-afternoon the fighting had intensified in all sectors. A desperate hand-to-hand struggle for St Laurent had ended with the 2nd Battalion, 916th Grenadiers, being pushed to the western suburbs

to which he had contributed so many years ago. Looking back over the intervening years, Severloh commented:

'I never saw a hero. A hero was someone who didn't get killed that day. I spent the war organising food and drink and shooting the bull with my superiors. I remember shortly before the invasion one officer said to us, "A stinking German corpse cannot save the Fatherland." Another told us to remain at our posts and die a hero's death. I remember telling Freking, "Wait and see, the enemy's not going to get that officer's stinking corpse." And they didn't!'

THE AUTHOR Paul Stelb is a military historian and militaria dealer. He and the publishers would like to thank Michael Passmore and Christopher Coxon, who are members of the WWI Battle Re-enactment Association, for their help in the preparation of this article.

THE 10TH MOUNTAIN DIVISION

After the entry of the United States into World War II, it soon became clear that the US ground forces had to be prepared to fight in a wide variety of terrains and climatic conditions.
With the assistance of the National Ski Patrol (NSP), the 87th Mountain Infantry Regiment was therefore raised and trained for specialised mountain operations. In June 1943, the US Army authorised the formation of the 10th Light Division (Pack, Alpine) using the 87th Infantry as its nucleus. The NSP recruited over 11,000 skiers, mountaineers and sportsmen for the new unit, bringing its total strength to more than 14,000.
After being taught skiing and rock climbing, the 'mountaineers' developed their tactical skills during a series of punishing winter exercises. Without doubt the best-trained US division, the unit joined the Fifth Army in the mountains of Northern Italy in January 1945.
Following the successful attack on Monte Belvedere, the 10th Mountain Division spearheaded the Allied drive through the Po Valley and on to Verona. By the time of the German surrender on 2 May 1945, the division had been in action for 114 days and lost more than 4000 killed and wounded. Even in this short time, however, the mountaineers had made a lasting impression upon their opponents.
The unit was de-activated on 1 December 1945.
Above: The shoulder patch of the 10th Mountain Division.

The storming of Riva Ridge in February 1945 was only the prelude to one of the US 10th Mountain Division's most memorable battles

BY DECEMBER 1944, the harsh winter conditions in the Apennine mountains of northern Italy had forced the Allied Fifth and Eighth Armies to suspend their attempts to smash through the German defences that barred their way into the Po Valley. Field Marshal Sir Harold Alexander, Supreme Allied Commander in the Mediterranean theatre, and Lieutenant-General Mark Clark, commanding the Allied armies in Italy, had agreed upon 1 April as the provisional date for an offensive that was designed to seize Bologna and encircle Field Marshal Albert Kesselring's Army Group C in the Po Valley. Major-General Lucian K. Truscott, commanding the US Fifth Army, selected Highway 64, running through the valley of the Reno River to Bologna, as the primary axis of advance for his army in the spring offensive. To secure a jumping-off point, Truscott decided upon a limited offensive, codenamed Operation Encore, to capture the high ground overlooking Highway 64.

Monte Belvedere is at the western end of a hill mass that extends east and northeast for a distance of eight miles, dominating Highway 64 about five miles to the east. The peaks and ridges in this region, averaging 3800ft in height, were the objective of the first phase of Operation Encore. Four infantry battalions of the German 232nd Grenadier Division were well dug in along this feature, and could call upon the fire of some 80 guns in support. The mountains' forward slopes, which descended towards the American positions across the Silla River, were heavily mined with a mixture of anti-tank and anti-personnel mines. Three American attacks on the formidable defences along the Monte Belvedere-Monte della Torraccia Ridge had already been repulsed, and it was with this in mind that Clark had requested that the 10th Mountain Division be sent to join the Fifth Army. Although the 10th had yet to see combat, it was acknowledged to be the best-trained division in the army, with a unique ability to operate in mountainous terrain under severe winter conditions.

Hays was confident in the ability of his superbly trained mountain troops to scale Riva Ridge

By 28 January 1945, only 10 days after its last units had arrived in Italy, the 10th Mountain Division was assembled in the Reno Valley, facing the German positions atop the Monte Belvedere-Monte della Torraccia Ridge. Studying the terrain, Major-General George Hays, the unit's commanding officer, soon came to the conclusion that the capture of Riva Ridge was the key to a successful assault on Monte Belvedere. Across the Dardagna River to the southwest of Monte Belvedere, Riva Ridge stretched for four miles; from Monte Spigolino in the south, to Pizzo di Campiano in the north. From their positions atop Riva Ridge, some of which were 6000ft above sea level, German observers would be able to direct devastating artillery fire onto the left flank and rear of any force that attempted to take Monte Belvedere.

Fifth Army Headquarters had already dismissed the capture of Riva Ridge as impossible – the side of the ridge facing the American positions was a cliff, rising in some places almost 1500ft above the valley

Below: Wearing his divisional insignia and combat infantryman's badge with obvious pride, this staff sergeant from the 10th Mountain Division carries one of the ice-axes favoured by America's alpine troops. Left: Engineers of the 10th Mountain Division take cover behind an M10 tank destroyer during the fighting in northern Italy. Below right: 'Mountaineers' advance northwards through the Apennines, having toppled the German defenders from their positions on the Monte Belvedere-Monte della Torraccia Ridge.

floor. But Hays was confident in the ability of his superbly trained mountain troops to scale Riva Ridge. Moreover, Hays reckoned, if Fifth Army Headquarters did not think the ridge could be taken by direct assault, neither would the Germans. Aware that its capture would secure the flank of the 10th Mountain Division for the main assault on Monte Belvedere, General Truscott approved Hays' plan.

In the first weeks of February, reconnaissance patrols probed forward across the Silla Valley in search of routes through the minefields on the forward slopes of Monte Belvedere. Other patrols, comprising some of the most highly trained mountain climbers in the division, searched for routes up the steep face of Riva Ridge. Although some of these recces were conducted on foot, skis and snowshoes often had to be employed in order to wade through snow that lay up to four feet deep in places. One patrol report concluded: 'Very high wind. Visibility poor. Snow is knee to waist-deep. Could not see Monte Spigolino. Had to dig footholds to timber. Crampons and ice-axes needed badly.'

At the beginning of February, the 1st Battalion and Company F of the 2nd Battalion, 86th Mountain Infantry Regiment, had been withdrawn from the line and transported to Lucca, near the Ligurian coast. For two punishing weeks, the 'mountaineers'

THE MOUNTAINEERS

10th Mountain Division
Operation Encore,
17-23 February 1945

By the bitterly cold winter of 1944, the advance of the Allied Fifth and Eighth Armies into northern Italy had bogged down on the slopes of the Apennine mountains. As the troops settled into their billets to sit out the worst of the weather, senior US and British commanders began to plan the next stage of the campaign: the seizure of Bologna and the clearing of the Po valley. Before an all-out offensive could be launched, however, it was decided to unleash the US 10th Mountain Division against a series of heavily defended ridges and mountains that dominated the main road between Rome and Bologna, Highway 64.

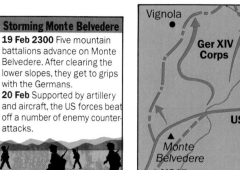

Storming Monte Belvedere

19 Feb 2300 Five mountain battalions advance on Monte Belvedere. After clearing the lower slopes, they get to grips with the Germans.
20 Feb Supported by artillery and aircraft, the US forces beat off a number of enemy counter-attacks.

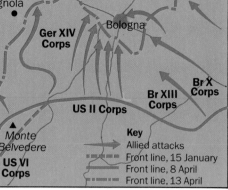

Vignola
Bologna
Ger XIV Corps
Br X Corps
Br XIII Corps
US II Corps
Monte Belvedere
US VI Corps

Key
→ Allied attacks
---- Front line, 15 January
— Front line, 8 April
···· Front line, 13 April

Climbing Riva Ridge

17 Feb Men of the 10th Division occupy Lizzano and Vidiciatico.
18 Feb 2300 The assault troops begin the climb up Riva Ridge.
19 Feb Despite heavy enemy fire, the ridge is secured. The main attack can now get underway.

Monte della Torraccia
Hill 1088
Valpiana Ridge
Fanano
Monte Golgolesco
Monte Castello
Pizzo di Campiano
Monte Belvedere
Riva Ridge
Florio
Polla
Monte Cappel Buso
1/87th
2/87th
1/85th
3/85th
3/86th
1st Div BEF
Gaggio-Montano
Monte Mancinello
Dardagna river
Quérciola
Company F, 2/86th
Vidiciatico
Lizzano
Silla river

The assault succeeds

21-22 Feb The fighting for the high ground continues.
23 Feb The Germans concede defeat and finally retreat.

Key
→ 10th Mountain Division attacks
---- Brazilian Expeditionary Force
⊣⊣⊣ German positions
→ German attacks

Above: Wounded personnel are evacuated from the slopes of Monte Serrasiccia during the fighting for Riva Ridge. This remarkable tramway was conceived and built by engineers of the 10th Mountain Division. The engineers continued their sterling efforts in support of the combat troops when the battlefield shifted to Monte della Torraccia, to the east. Before the main assault began, engineers conducted an extensive sweep of the area to neutralise the German mines that littered the forward slopes of the division's objective (above right). Following the successful attack on Monte Belvedere, the 10th Mountain Division led the Allied drive through the Po Valley and into Verona. Main picture: Elements of the division's self-propelled artillery shell German positions.

refreshed their rock-climbing skills in a marble quarry. These crack troops had been selected by General Hays to scale Riva Ridge on the night of 18 February, 24 hours before the division's attack on Monte Belvedere. Four routes had been identified up the face of Riva Ridge, and the three rifle companies of the 1st Battalion were each assigned a route up the eastern half of the ridge. To the west, Company F would make an assault on Monte Mancinello while the remainder of the 2nd Battalion covered its flank. The right flank of the assault force would be covered by Company E, the heavy weapons company of the 1st Battalion.

During the night of 17 February the 1st and 2nd Battalions of the 86th Infantry moved quietly into the Silla Valley and leaguered in the villages of Vidiciatico and Lizzano. Throughout the following day, the men carefully prepared their equipment for the assault on Riva Ridge. Weapons were cleaned and oiled, pitons checked and ropes, made of newly-developed lightweight nylon, inspected and coiled. The veteran climbers selected to lead the assault spent hours studying the now-familiar cliff face through their binoculars. In addition to his weapons and ammunition, each man would carry two days of cold rations, a full canteen, entrenching tools, two blankets and an extra sweater.

At 1930, in complete darkness, the mountain troops of the 86th Infantry, jackets tied about their waists, moved out in silence from their temporary barracks. Dividing into four columns, they moved towards the towering snow-and ice-covered mass of Riva Ridge. Crossing the shallow water of the Dardagna River, the lead troops soon began crunching through the snow at the foot of the ridge, reaching the base of the cliff at 2300 hours. Peering skywards up the 1000ft cliff face, the lead climbers carefully selected their first toe-holds for the ascent. Where none could be found, the climbers used muffled hammers to drive in pitons, secured snap links to them and then fastened ropes.

Slowly, painfully, the four files inched their way up the cold, icy rock face. Atop the ridge, totally una-

ware of any American activity, German troops of the 232nd Fusilier Battalion were relieving the 2nd Battalion, 1044th Infantry Regiment. By the first streaks of dawn on 19 February, most of the assault force had reached the top of the cliff undetected. The first German sentry to see the mountaineers appear through the haze stared in utter disbelief before raising his hands in a gesture of surrender. Moving stealthily towards a bunker, one member of the assault force tripped over a concealed alarm wire. Now that the need for silence had disappeared, the Americans set about their task with lightning speed. One by one, the bunkers were cleared with a few well-placed hand grenades. Surprise had been all but complete – with only 34 casualties, including seven dead, the four companies of the 86th were now in possession of Riva Ridge. It took the Germans almost an hour to appreciate what had happened and direct their artillery fire onto the ridge, but the enemy guns were quickly silenced by American counter-battery fire.

After considering the various options, Hays decided to launch a surprise night attack

With the 10th Mountain Division's left flank now secure, General Hays could devote his full attention to the capture of Monte Belvedere. After considering the various options, Hays decided to launch a surprise night attack without any artillery preparation. Although this seemed a hazardous plan with an untried division, the losses anticipated during a daylight assault across the barren approaches to Belvedere were considered unacceptable.

Five battalions were to participate in the first phase of the attack. On the left, the 1st and 2nd Battalions of the 87th Mountain Infantry Regiment would launch their attack towards Valpiana Ridge, a spur to the northwest of Monte Belvedere. In the centre, the 1st and 3rd Battalions of the 85th Mountain Infantry Regiment would attack Monte Belvedere and Monte Gorgolesco. On the right flank, the 3rd

mediate response from the artillery under Hays' command. The air became thick with the sound of exploding shells and mortars.

Company F, 2/87th, advancing on Company G's left flank, came under fire at 0020 hours. By 0220 the company was pinned down by heavy machine-gun and mortar fire from the village of Florio. One hour later, supporting artillery fire was directed onto the enemy strongpoints with the desired effect. At dawn on 20 February the company's three platoons stormed the village. Three German defenders were killed and 55 captured. By mid-morning the 2nd Battalion of the 87th Infantry had sent more than 160 prisoners to the regimental command post. Interrogation soon revealed the extent of the battalions' success. Their objective, Valpiana Ridge, had been held by four companies of the German 1044th Infantry Regiment, each between 80 and 100 strong. Despite being reinforced by an anti-tank company, the defenders had been virtually wiped out.

All along the line of the Riva Ridge, the mountaineers were meeting with similar success. The 85th Infantry had advanced to within 300yds of the summit of Monte Belvedere before encountering any opposition. Shortly after 0600 the 3rd Battalion was in possession of Monte Belvedere, and an hour later the 1st Battalion had driven the last German defenders off Monte Gorgolesco. The 3rd Battalion of the 86th Infantry met comparatively little opposition and secured its objective soon after 0600 hours.

Battalion of the 86th Infantry would attack the southern slope of Monte Gorgolesco. The 1st Division of the Brazilian Expeditionary Force would secure the 10th Division's right flank by assaulting Monte Castello, to the southeast of Monte della Torraccia.

To augment the 10th Division's three battalions of 75mm pack howitzers, General Hays was assigned four batteries of 105mm howitzers from the 175th and 1125th Field Artillery Battalions, the 4.2in mortars of the 84th Chemical Mortar Battalion, the 701st and 894th Tank Destroyer Battalions and the 751st Tank Battalions. Although the tanks and tank destroyers could only be brought to within a mile of the Belvedere-Torraccia Ridge, they could be counted upon to provide invaluable direct fire support.

Mortar and artillery fire was now pounding the valley floor and the forward slopes of the ridge

After an approach march of several miles, at 2300 on 19 February the leading elements of the five mountain battalions simultaneously crossed the line of departure and silently advanced towards the minefields that protected the Belvedere Ridge. The advance went undetected until shortly after midnight, when two enemy sub-machine guns opened fire on Company B of the 1st Battalion, 87th Mountain Infantry Regiment (1/87th). Minutes later, further to the west, the troops of Company G, 2/87th, detonated a series of mines outside the village of Polla. Although the 1st Platoon suffered several casualties, it overran a mortar position on the outskirts of the village. Faced with intense smallarms fire raking the ground around them, the men of the 2nd and 3rd Platoons were forced to take shelter in the midst of the minefield and await the arrival of engineers from the 126th Mountain Engineer Battalion. German mortar and artillery fire was now pounding the valley floor and the forward slopes of the ridge, provoking an im-

Right: Out on patrol in the snow-covered Reno Valley, searching for routes up the forward slopes of Riva Ridge. **Above:** Troops of Battery B, 616th Pack Howitzer Battalion, 10th Mountain Division, hump 75mm ammunition over rough terrain during the shelling of the ridge. **Above, far right:** Men of the 110th Signals Company set up communications wire as the division advances towards Camidello, south of Verona.

When the early morning sun dispelled the darkness on 20 February, a new element entered the battle. At 0700, Spitfires and P-47 Thunderbolts of the 22nd Tactical Air Command arrived in the skies over Monte Belvedere. Directed by forward air controllers, codenamed 'Rover Joes', the fighters screamed out of the sky to rocket, bomb and strafe German positions. This demonstration of Allied airpower further raised the morale of the mountaineers – during the first two days of Encore, the fighters flew 412 sorties in support of the infantry attacks.

As German mortar and artillery fire intensified, the Americans began mopping up isolated pockets of enemy resistance and prepared for the inevitable counter-attacks. At 0730 and 0930 Company B, 1/87th, fought hard to break up a series of small counter-attacks against its position on Valpiana Ridge. On the right flank, the 3rd Battalion, 87th Infantry, beat off

George Price Hays (1892-1978) was commissioned as a second lieutenant in the United States Army Field Artillery in 1917 and fought with the 3rd Infantry Division at Château-Thierry, Champagne-Marne, Aisne-Marne, St Mihiel and Meuse-Argonne. Hays earned the Congressional Medal of Honor in the second battle of the Marne when, by directing the Allied artillery fire, he saved the 3rd Division from being overrun.

Between the wars, Hays held various command and staff appointments both in the United States and the Philippines. In 1942 he joined the 2nd Infantry Division as commanding general of artillery. Apart from a temporary assignment commanding the artillery of the 34th Division during the assault on Cassino, Hays remained with the 2nd Infantry Division until November 1944.

Major-General Hays was then selected by General George Marshall, the American Chief of Staff, to command the 10th Mountain Division and lead it during its combat debut in Italy. His dynamic leadership acted as a tremendous fillip to the division's morale, and inspired General Truscott to describe Hays as one of his ablest combat commanders.

Before retiring in 1953, General Hays served as the Commanding General of the Military Government for Germany, and as Commander of US Forces in Austria. A veteran of two world wars, General Hays was awarded the Distinguished Service Medal, the Silver Star, the Legion of Merit, the British Companion of the Order of Bath, and the French Légion d'Honneur and Croix de Guerre.

Officer, US 10th Mountain Division, Italy 1945

Wearing wool trousers and khaki temperate shirt, this officer carries an M1 semi-automatic rifle, and a cotton bandolier for storing smallarms ammunition. Around his waist is an officer's pistol web belt.

19

three platoon and company-sized counter-attacks. Supported by artillery, tank fire and fighter-bombers, the battle-hardened troops of the 10th Mountain Division drove off every counter-attack that the enemy was able to mount. As darkness drew in, the 2nd Battalion of the 85th Infantry passed through the 1st Battalion and was in possession of Cappela di Ronchidos by 2030.

Early on the morning of 21 February a battalion of the German 714th Jäger Division, reinforced by elements of the 1043rd Infantry Regiment, launched a counter-attack on the positions of the 3rd Battalion, 85th Infantry, on Monte Belvedere. Aided by the suppressive firepower of Allied artillery and air support, the defenders managed to beat off the attack, inflicting heavy casualties on the enemy. That same afternoon, the 2nd Battalion of the 85th Infantry began pushing towards Monte della Torraccia – the last bastion of German resistance. The Germans put up a stiff fight, however, and progress was painfully slow. By 22 February the battalion was still 400yds short of the crest and reduced to an effective strength of only 400 men. General Hays therefore ordered the men of the 3rd Battalion, 86th Infantry, to relieve their embattled compatriots. .

On the morning of 23 February, while Spitfires and Thunderbolts bombed and rocketed enemy positions and gun batteries, the 3rd Battalion launched its assault. Company I fought its way to the top of the crest within one hour, and Monte della Torraccia was cleared of German defenders by the early afternoon. The 4th Independent Mountain Battalion, part of the 232nd Division's reserve, then launched a vigorous counter-attack. Although 1000 artillery and mortar shells landed atop the mountain in support of the enemy attack, the German Bergtruppen were repulsed by the American mountain troops. The battle for Monte Belvedere was over.

Below, inset: Medics of the 10th Mountain Division carry wounded men to the field hospital, located behind the unit's forward positions.
Below: Watched by men from one of the division's tank destroyer battalions, a mountaineer (foreground) marches German prisoners to a stockade in northern Italy. Each prisoner has been tagged with a white label, indicating his rank and where he was captured. This procedure was instituted to ease the burden of the Allied Counter-Intelligence Corps (CIC). Despite being in action for only 114 days, the 10th Mountain Division built up a formidable reputation among its opponents. General von Senger, the commander of the XIV Panzer Corps, later described it as the best division he had faced on any front.

In five days of furious combat, the men of the 10th Mountain Division had captured the seemingly impregnable Riva Ridge and the eight-mile long Monte Belvedere-Monte della Torraccia Ridge. Field Marshal Kesselring had believed that Allied units were incapable of attacking this formidable defensive position under such severe weather conditions. Although the battle cost the division 850 casualties, including 195 killed, the tenacity of the American mountaineers had proved Kesselring wrong. The success of the operation had been entirely dependent upon the bold and imaginative use of an elite division trained to fight and win in even the most unfavourable of terrains. The achievements of the 10th Mountain Division in its first action were summed up in a commendation issued to the troops:

'You accomplished all of your assigned mission with magnificent dash and determination. You caught the enemy completely by surprise by your movement at night up precipitous slopes through his heavily mined areas, and by your destruction of his dug-outs and bunkers. You overran and defeated elements of eight different enemy battalions, from which you captured approximately 400 prisoners of war...As your division commander, I am very proud of you and salute your courage, determination, fighting spirit and the professional workmanship you have displayed in all your actions.'

THE AUTHOR Ian Kemp, a former member of both the British and the Canadian Armies, graduated from the Department of War Studies, King's College London, and is a researcher at the Royal United Services Institute for Defence Studies.

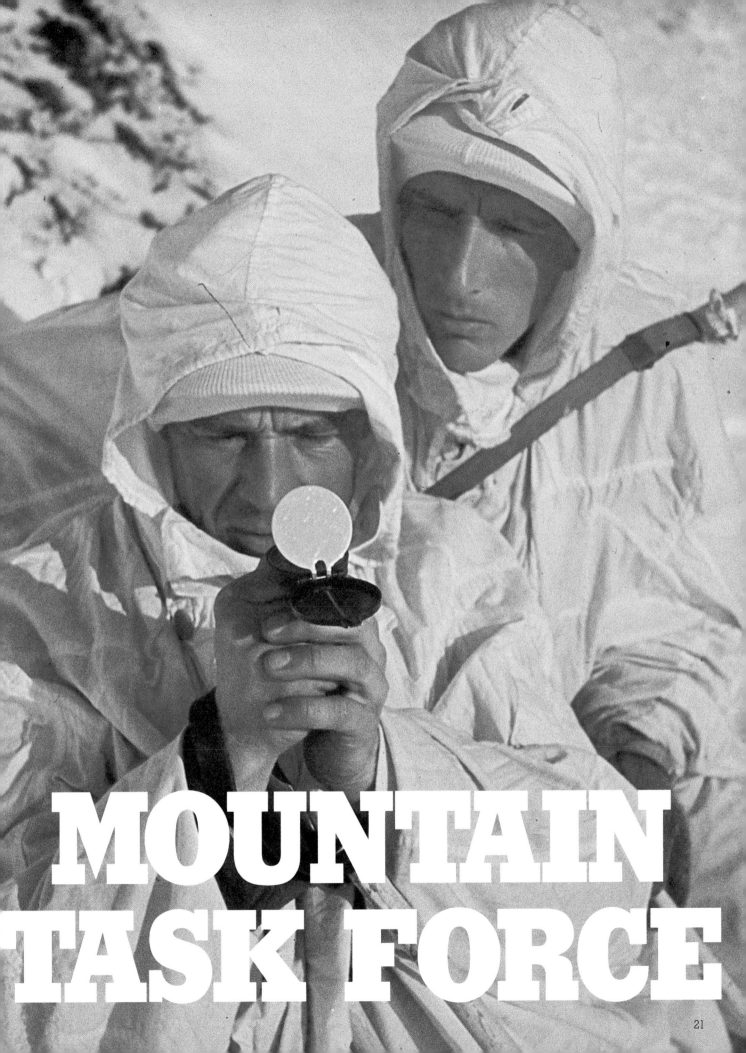

MOUNTAIN TASK FORCE

MOUNTAIN TROOPS

The 3rd Mountain Division was raised in 1938 from the 5th and 7th Mountain Divisions of the Austrian Army, following the German annexation of Austria. The division had two infantry regiments (each of three battalions), an artillery regiment and full supporting arms and services. Commanded by Major-General Eduard Dietl (until 1941) the division was well versed in mountain warfare, having been rigorously trained by Dietl himself. Their martial skills were first put to the test in September 1939 when they distinguished themselves during the Polish Campaign. In mountainous or rough country their flexible structure and excellent training made them formidable opponents, though lack of heavy artillery made them vulnerable in open terrain. The artillery regiment was equipped with 16 7.5cm guns, eight 10.5cm guns, and eight 15cm guns. Following Hitler's invasion of the Soviet Union in June 1941, the 3rd Mountain Divison was transported to the Baltic states and, for the rest of the war fought on the Eastern Front. Alongside other mountain units, it was deployed in the northern sector and took part in the attack on Murmansk, and earned distinction in the Zaporozhe area in the autumn of 1943.
Above: The Narvik Shield, awarded to commemorate the campaign, was worn on the left upper sleeve.

In April 1940 Hitler moved into Norway where the 3rd Gebirgsjäger (Mountain) Division was stretched to the limit around the port of Narvik

AT DAWN on the morning of 9 April, the 10 destroyers of German Naval Group 1 entered the fjord leading to the port of Narvik. On board, packed into every available space, were the soldiers of the 139th Mountain Infantry Regiment and the headquarters of the 3rd Gebirgsjäger (Mountain) Division. The sea journey had been through appalling weather conditions: snow blizzards and rough seas had resulted in men, equipment and weapons being washed overboard, with the survivors suffering from exposure and sea-sickness. The Germans planned to surprise and overwhelm the Norwegian naval and military defences at Narvik, and this was largely achieved within a few hours. Two Norwegian naval vessels which opened fire on the German destroyers were blown out of the water, and in the ensuing confusion the mountain troops were successfully landed, whilst the divisional commander, Major-General Eduard Dietl, negotiated the surrender of the Norwegian forces with the local commander, Colonel Konrad Sundlo. By 0810 Dietl was able to report to Oberkommando der Wehrmacht (OKW) that Narvik was in German hands. However, over 200 Norwegian soldiers refused to surrender and marched out of the town determined to continue the fight. This made the position of Dietl's mountain troops extremely precarious; they had no heavy weapons and were cut off by land and sea from the main German forces that had been landed at various points on the Norwegian coast further to the south.

Dietl established a remarkable rapport with his men and demanded the highest possible standards

Nearly all the German ground forces allocated to Norway were second-line units, apart from the 3rd Mountain Division. The division was split in two, with the divisional headquarters and the 139th Mountain Infantry Regiment being landed at Narvik while the 138th Mountain Infantry Regiment was landed at Trondheim. Eduard Dietl, the divisional commander was a professional soldier who had fought in the infantry during World War I. He was 50 in 1940, and had commanded the division since it was raised. Dietl was a Bavarian and a Nazi sympathiser who was regarded with particular favour by Hitler. An expert skier and a devotee of mountain climbing, Dietl had become a pioneer of mountain training in the German Army. A hard, tough soldier, he established a remarkable rapport with his men and demanded the highest possible standards from all ranks. He believed in realistic training, and before the war the 3rd Mountain Division had practised for all possible types of warfare. Dietl's military philosophy can be summarised in his own words:

'Soldiers must be led by the heart. Only then are they committed. All else is in vain. He who has the soldier's heart can beard the devil in hell with him. There has to be drill – that's part of the trade. But mere drill alone is nothing. It is worse than no drill. Drilling is easy. Leading is hard. Leadership calls for two separate things. The first, certainly, is – live with the man. Wish to have nothing but what he has. Go with him, listen to him, understand him, help him in tough places. But the second is – be

Page 21 : Two recruits of the Wehrmacht's crack 3rd Mountain Division are put through their paces during training; navigational skills were vital in their often bleak combat environment.
Left: One of the skills that set the mountain men apart from other troops was abseiling.
Below: The scene in Narvik harbour after the Royal Navy's attack in the early part of the Norwegian campaign. Right: Major-General Dietl (right), commander of the division, surveys forward defensive positions during his masterful defence of Narvik.

better than the man. Never forgive yourself anything. Always know what you, as a leader, have to do. Be hard, if necessary, demand the utmost, but first do the utmost yourself.'

The operational orders for Dietl, leading the 139th Regiment, were to seize and hold Narvik and secure the vital iron-ore railway to Sweden. Then, after the expected reinforcements reached him from the south, Dietl was to secure the area between the Swedish frontier and the coast. But on 9 April, after a reasonably successful landing, Dietl was concerned to establish his position at Narvik rather than consider the more grandiose elements of his operational orders. His mountain troops were isolated in the Arctic mountains and fjords, short of heavy weapons and ammunition. The town of Narvik itself was particularly vulnerable, occupying a small area of reasonably level ground at the tip of a peninsula, nine kilometres long and four kilometres wide, flanked to the north and south by fjords. The railway ran eastwards, following the shore of the southern fjord along a narrow shelf, through tunnels cut into the mountains further inland. From the town, the Arctic wilderness stretched away in all directions, a mass of hills, depressions and odd-shaped plateaus. In April, winter still howled around Narvik, with snow drifts blanketing the town centre and inland valleys. Later, the cold spring rains were to make life wretched for all combatants at Narvik.

Dietl quickly set about the defence of Narvik, establishing divisional headquarters in the Hotel Royal. Although German naval losses at Narvik over the next week meant that sea communications were cut, the loss of all the German destroyers reinforced the 4600 men of the 3rd Mountain Division with 2600 sailors armed with an assortment of weapons captured from the Norwegian 6th Division. Dietl deployed his mountain troops and sailors to defend the approaches to Narvik and the town itself. Two battalions of mountain troops were positioned 27km north of Narvik, while the remaining battalion, reinforced

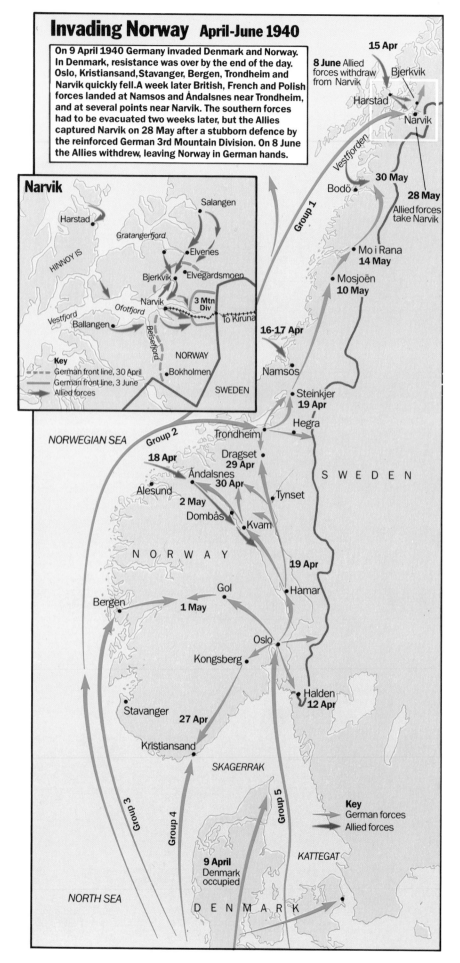

Invading Norway April-June 1940

On 9 April 1940 Germany invaded Denmark and Norway. In Denmark, resistance was over by the end of the day. Oslo, Kristiansand, Stavanger, Bergen, Trondheim and Narvik quickly fell. A week later British, French and Polish forces landed at Namsos and Åndalsnes near Trondheim, and at several points near Narvik. The southern forces had to be evacuated two weeks later, but the Allies captured Narvik on 28 May after a stubborn defence by the reinforced German 3rd Mountain Division. On 8 June the Allies withdrew, leaving Norway in German hands.

Narvik

Key
- - - German front line, 30 April
- — German front line, 3 June
- → Allied forces

15 Apr

8 June Allied forces withdraw from Narvik

Bjerkvik

Harstad

Narvik

Vestfjorden

Group 1

30 May

Bodö

28 May
Allied forces take Narvik

Mo i Rana
14 May

Mosjöen
10 May

16-17 Apr

Namsos

Steinkjer
19 Apr

Hegra

NORWEGIAN SEA

Group 2

Trondheim

18 Apr

Dragset
29 Apr

Åndalsnes
30 Apr

Alesund

2 May

Dombås

Kvam

Tynset

SWEDEN

N O R W A Y

19 Apr

Gol

Hamar

Bergen

1 May

Oslo

Kongsberg

Halden
12 Apr

Stavanger

27 Apr

Kristiansand

SKAGERRAK

Group 5

Group 3

Group 4

Key
- → German forces
- → Allied forces

9 April
Denmark occupied

KATTEGAT

NORTH SEA

D E N M A R K

Narvik inset map:

Harstad

Salangen

Gratangerfjord

Elvenes

HINNOY IS

Bjerkvik

Elvegardsmoen

Narvik

Ofotfjord

3 Mtn Div

To Kiruna

Vestfjord

Ballangen

Beisfjord

NORWAY

Bokholmen

SWEDEN

---

by the naval crews, took up positions inside Narvik and around the fjords.

Hitler and OKW were growing increasingly concerned about the position of Dietl's force at Narvik; they realised that the Royal Navy had blockaded the coast and would prevent German reinforcements arriving by sea. Furthermore, the Allies were in a position to land troops and overwhelm Dietl's isolated force. The Germans attempted to reinforce Narvik by air, and although an artillery battery was successfully landed on a frozen lake on 13 April, the thawing of the ice stopped the aircraft from taking off again. Bad weather then prevented the Luftwaffe from making any substantial re-supply by air until the middle of May.

One of the 3rd Mountain Division's main advantages was that the Swedish frontier was only some 30km to the east along the railway. So, on 16 April, units of the 3rd Mountain Division attacked along the railway, dispersing the Norwegian troops and reaching the Swedish frontier. The control of this position by Dietl enabled Berlin to exert considerable pressure upon the Swedish government, forcing them to compromise their neutrality; although the Swedes refused to allow the Germans right of passage for weapons, equipment and armed troops, they did allow them to transport medical personnel and supplies of clothing and rations through to the men at Narvik. However, this supply line along the railway was vulnerable where it ran along the fjord because ships of the Royal Navy were able to maintain an effective naval bombardment.

On 24 April a British battleship, two cruisers and six destroyers shelled Narvik for three hours

Whilst Dietl was deploying his forces, the Allies were beginning to react to events around Narvik. Churchill, the First Lord of the Admiralty, realised the importance of the town and urged an immediate landing by British military forces. By 14 April a British advance landing party had arrived off Narvik, but the commander, Major-General Mackesy, loath to commit his force against an unknown number of Germans, urged the Royal Navy to carry out a naval bombardment to induce the Germans to surrender. On 24 April a British battleship, two cruisers and six destroyers shelled Narvik for three hours. The Germans expected a landing under cover of the bombardment and Dietl informed OKW that in the event of his being unable to hold the town, then he would withdraw eastwards along the railway. When no landing was undertaken, Dietl withdrew non-essential forces from Narvik and moved his command post to a railway station at the eastern end of the southern fjord.

There is no doubt that in the first two weeks the Allies greatly overestimated German strength at Narvik and were far too cautious; an immediate landing would have forced Dietl to abandon the town. Instead, the British force established its headquarters at Harstad on Hinnöy Island to the west, and gradually more Allied troops were formed into two groups, one to the north and one to the south of Narvik. There were three battalions of British infantry, four of Norwegian and three battalions of French Chasseurs Alpins. By the beginning of May this force had been reinforced with two battalions of the French Foreign Legion, a Polish brigade and additional Norwegian units. The Royal Navy had overwhelming superiority at sea with a force that included an aircraft carrier and a battleship. However, the Allies, lacking adequate AA guns, were extremely vulnerable to Luftwaffe attacks supporting Dietl's force.

Despite the Royal Navy's numerical advantage, Dietl was convinced that the Allies would not undertake a major operation against Narvik itself until they had planned everything down to the last detail, and would probably wait until the snows had melted. He was determined to hold his defensive position to the north of the town with two battalions of mountain infantry; leaving the town itself to the remaining battalion.

In the north, Dietl's two battalions found themselves being slowly pushed south and in continuous danger of being outflanked. Then, on 13 May, Dietl was faced with the landing of two French battalions with light tanks towards the top of the northern fjord, threatening to attack his main defensive position from behind. Short of mountain troops, Dietl had been forced to deploy his ad hoc naval personnel to hold this area, and badly shaken by a naval bombardment, they hastily withdrew in the face of the French troops. Confronted with this dangerous threat to the rear of his northern defensive position and with renewed Norwegian attacks against the right flank, Dietl was forced to withdraw his two mountain infantry battalions south to a position near the Storebalak. Dietl reported to German headquarters in southern Norway that his position was critical and required immediate reinforcements. He intended, if the Allied offensive continued, to abandon Narvik and to hold a bridgehead on the railway. But, unless he received reinforcements, he would have to consider taking his force into Sweden, preferring internment to the alternative of Allied capture.

Right: Dietl's mountain troops undergo ski training. Warfare in the bleak, inhospitable conditions of the frozen north required very particular skills and immense reserves of energy. Dietl could not emphasise enough the importance of realistic training and it was through this and his rigorous leadership that his men were able to hold out when they ran short of food and winter clothing. Below: Mobility is all important to mountain warfare and troops were trained to use inflatable boats for crossing fjords.

The German position at Narvik was precarious, despite the fact that since 4 May the German 2nd Mountain Division had begun to move north from Trondheim and the German offensive against France had been launched on 10 May. The 2nd Mountain Division was still 280km to the south of Narvik and it was too early to estimate the effect that the offensive would have on the Narvik sector.

The three battalions of mountain infantry were reaching the limit of their endurance

On 14 May German headquarters in southern Norway flew in a token force of 66 paratroopers as reinforcements for Narvik, and during the remainder of the month and early June the 137th Mountain Infantry Regiment was dropped over the Narvik perimeter (after hasty parachute training), reinforcing Dietl with some 1050 men. Of Dietl's original force, the naval personnel were not trained or prepared for any lengthy period of combat, and could not be expected to hold ground against determined Allied attacks. The three battalions of mountain infantry were reaching the limit of their endurance, and the two battalions fighting to the north of Narvik had suffered badly as a result of the appalling

THE GERMAN PLAN

The German operational plan to occupy Norway and Denmark, codenamed *Weserübung,* was a gamble. Hitler was concerned in the winter of 1939/40 that the British could interrupt his ore supplies from Sweden, which travelled through Narvik and along the Norwegian coast. The seizure of Norway by Germany would also deny the Allies control of trade routes in the area and the opportunity to occupy bases from which to threaten German control in the Baltic. Hitler calculated that if sufficient German forces could be landed by sea and air at strategic points along the Norwegian coast, they could overwhelm the limited Norwegian defences without any serious fighting, and present Britain with a fait accompli. But even if the German forces could surprise the Norwegians, the plan was still a gamble; British naval superiority in the area could prevent the initial landings or blockade the coast. To execute *Weserübung,* the German Navy planned to land troops in 11 groups from naval and merchant vessels at key points, including the widely separated cities of Oslo, Kristiansand, Bergen, Trondheim and the northern port of Narvik. The majority of these forces would be landed in southern Norway, but in the event of serious Norwegian resistance further north around Trondheim and Narvik, they would have to move north as reinforcements.

25

Below: Crouching behind the scant protection offered by a convenient snowdrift, two Gebirgsjägern await a renewal of the Allied offensive to capture Narvik. Although the town fell after a protracted fight, Dietl was able to reoccupy the port when the enemy's expeditionary force withdrew on 8 June. Bottom: Mountain troops, accompanied by mule-packed mountain guns, move down from one of Norway's snow-covered valleys.

weather conditions – driving sleet and snow, and extreme cold and exposure. When forced to withdraw, they had abandoned reserves of clothing and food.

The Germans were so exhausted that they could only be persuaded to move from one position to another with great difficulty, and frequently fell asleep during machine-gun and shell fire. But the conditions were equally bad for the Allies, and this, combined with the difficult nature of the terrain and poor communications, meant that they were unable to exploit the German weaknesses.

On 21 May Dietl calculated that an Allied breakthrough in the north was imminent, and to prevent the collapse of his position, he withdrew to a shortened line on 22 May, based on the Swedish border at Björnfell. Ironically, just at the moment when Dietl was preparing for a final stand before abandoning Narvik and withdrawing into Sweden, the Allies were preparing to take Narvik in order to destroy the

port and then cover the evacuation of the landing force. The final Allied assault on Narvik consisted of an amphibious landing across the Rombaks fjord by two French battalions and one Norwegian battalion, whilst the Poles attacked from the south against Ankenes; thirteen Allied battalions were ranged against ten German battalions. The French and Norwegians would keep up the pressure further to the north, and after the capture of Narvik a further attack from the south would cut the railway line.

The attack began on 27 May, and although the French and Norwegians landed successfully to the east of Narvik, the 2nd battalion of the 139th Mountain Infantry Regiment, holding the high ground to the south, succeeded in defending the shore line long enough for the Germans to evacuate Narvik. After the Allies had captured Narvik the French 1st Chasseur Division began to advance eastwards along the Kiruna railway, and by 1 June Dietl was faced with concentric Allied attacks from north and south pushing his battered troops towards the Swedish frontier. Bad weather had prevented the Luftwaffe from making air drops, and supplies of ammunition and food were short.

Short of food, ammunition and winter clothing, usually very tired, they kept on fighting

Dietl was faced with defeat knowing that despite tremendous efforts it would be unlikely that German troops advancing from the south could reach him in time. But unbeknown to Dietl, the Allies had already begun to pull out of Narvik. Between the end of May and the 8 June some 15,000 troops and most of their guns and tanks were successfully evacuated. Almost at the time when Dietl was thinking of making the final move into Sweden, on the late afternoon of 8 June, he discovered that the Allies had gone. The surprised Germans marched back into an abandoned Narvik in triumph, and on 9 June the Norwegian High Command surrendered.

Nazi propaganda dubbed Dietl 'the Hero of Narvik'. A grateful Führer had already awarded him the Knight's Cross on 9 May, and on 19 July he was awarded the Oak Leaves and promoted to the rank of full general. Later, Dietl was to be awarded posthumously the Swords to the Knight's Cross following his death in an air crash in July 1944, whilst commanding the Twentieth Mountain Army in Lapland.

The Germans were fortunate at Narvik that the Allies were slow to react, lacking a decisive overall commander and a properly thought out plan. For its part, the 3rd Mountain Division had not expected to face a long, drawn-out campaign lasting two months under appalling weather conditions. It was a grim experience, particularly for the two battalions to the north of Narvik. Short of food, ammunition and winter clothing, usually very tired, they kept on fighting because of their esprit de corps, the personal leadership of officers and NCOs, and the forceful personality of Dietl their divisional commander. Although the German force at Narvik included naval personnel and paratroopers, the mountain troops of the 3rd Mountain Division bore the brunt of the fighting.

THE AUTHOR Keith Simpson is a senior lecturer in War Studies and International Affairs at Sandhurst. He is a member of the Royal United Services Institue and the International Institute for Strategic Studies. He has a special interest in modern warfare and has just completed a book on the German Army.

A key objective in the 1948 Arab-Israeli War, the town of Safed in central Galilee was secured by a vastly outnumbered Palmach force

IT WAS ON 12 April 1948, one month before the British Mandate over Palestine was due to end, that Major-General Sir Hugh Stockwell, GOC Northern Palestine, called a meeting with a group of Arabs from the Galilee town of Safed. On the agenda was Stockwell's plan for an orderly British withdrawal from Safed and its environs. He was worried that his forces would become embroiled in the fighting between the town's 12,000 Arabs and its minority of 1500 Jews, and for this reason he was prepared to turn over the town's strongpoints to the Arabs when the British withdrew. Stating that the Jews were outnumbered and could not possibly win any battle for the town, Stockwell gave his conditions to their agreement: the Arabs were not to attack the Jewish quarter until after the British withdrawal; they were to allow the British to intervene if the Jews requested it; and they were to treat the Jews humanely when the town was secured.

Right: During the period of the British Mandate to Palestine, the Haganah, and its strike element the Palmach, was an illegal organisation. Recruits were between 17 and 25 years old, although the services of former members of the Jewish Infantry Brigade Group in World War II were also highly valued.

TRIUMPH OF THE PALMACH

Operation Yiftach
Safed, May 1948

Ein Zeitun
Biriya
Kiryat Sarah
Mt Canaan ▲
Shalva House
Teggart Fortress
Police Station
The Citadel
Govt House
Safed

Key
→ Palmach attacks
Palmach reinforcements
→ Arab withdrawal
Jewish quarter
Arab quarter

MEDITERRANEAN SEA
Safed •
Sea of Galilee
Haifa •
SYRIA
ISRAEL
River Jordan
Tel Aviv •
Jerusalem •

The Haganah's success in Palestine owed much to its highly mobile Palmach strike units and its excellent communications network. Left: Ordered into battle, a half-track speeds Palmach men to the defence of a Jewish *kibbutz*. Below left: Volunteers in Tel Aviv man an intelligence post, receiving combat reports from outlying settlements under siege. Right: Jeep-mounted Palmach members tear into an Arab village. The jeep in front is covered by a machine gun in the following vehicle.

Two days later, the colonel commanding the British forces in the area met the leaders of Safed's Haganah unit (the Haganah was the Jewish community's self-defence arm and was expected to provide the nucleus of the future Jewish state's armed forces). Apprising them of the plan, he urged the 375 Haganah fighters in Safed to leave the town, allowing the town's population to come to a peaceful agreement on their own. Although they were told that the Arabs would gain control of the town in two hours, the Haganah leaders refused to evacuate either their men or the women and children of Safed from the town.

The highest town in Palestine, Safed is built around the slopes of a hill dominating the main roads of central Galilee. The town itself is dominated by a shrub-covered, even plateau about 150yds long, and on the sides of the hill are spread the Jewish and Arab quarters. The Jewish quarter was surrounded on three sides by Arab neighbourhoods, and below it lay the Arab villages of Ein Zeitun and Biriya.

The end of the Mandate was expected to bring in its wake an invasion by the regular armies of the neighbouring states. Faced with this threat, and that of the 2000 Arab nationalist fighters in the Safed area,

The Palmach, or Plugot Mahatz (Strike Companies), was set up by the Haganah in May 1941, when it looked as if there was nothing to stop the Afrika Korps from overcoming the Allies in the Western Desert and going on to invade Palestine. At the time of its founding, the primary aims were to protect Jewish towns and settlements against attacks and harassment from Arab gangs when the British withdrew, and to form the core of a guerrilla organisation against the occupying Axis troops. When the threat of an Axis invasion receded, the Palmach became the striking arm of the Haganah. At the same time, men from the organisation volunteered to support the Allies, and two Palmach companies spearheaded the Allied invasion of Lebanon and Syria.

Palmach members served full-time in small groups, scattered throughout Palestine. They were trained to fight as small units and emphasis was placed on leadership and initiative. They were especially adept at nightfighting. By 1947 the Palmach had become an elite force whose troops were to be found on virtually every front. It contributed three brigades (6000 men) to the Jewish forces that fought off the Arab invasion in 1948.

The Palmach was officially disbanded later in that year. It left a lasting impression on the Israeli Defence Forces, and many fundamental IDF military doctrines originated in the Palmach. Although many of its senior officers left the army following the end of the 1948-49 War, most of those who remained reached high rank, including, on five occasions, that of Chief of Staff of the IDF.

the Jews succeeded in establishing control of large areas of Arab land, covering most of the territory allocated to the Jewish state under the UN partition plan. They did not control Safed, however, and upon this town hung the fate of the expected invasion of the Galilee.

On 16 April the British evacuated Safed. With all the town's strongpoints now in Arab hands, the Jewish quarter was immediately placed under siege. Appeals for urgent help were sent from the heads of Jewish settlements all over northern Palestine to the Haganah headquarters in Tel Aviv. A plan for expelling enemy forces from Galilee, after which control would be assumed of the major transport arteries and the region prepared for the expected invasion, was drawn up by Yigal Feicovitch-Allon, the 29-year old head of the Palmach (the Haganah's elite striking arm). Allon's plan was approved and he was appointed commander; the mission was named Operation Yiftach, after the initials of its leader and the settlement of Tel Chai.

At Allon's disposal were all the Haganah forces in Galilee, as well as a new Palmach Brigade, also named Yiftach, formed from the Palmach's 1st and 3rd Battalions, which had already been fighting in the area. He decided that Safed, because of its strategic importance, would be the prime target of the operation. Not only would the Arab siege have to be broken, but the town itself would have to be captured before the Arab invasion. Operation Yiftach was now a race against time.

Almost as soon as the British departed, the Arabs mortared the Jewish quarter, and then attacked it on foot. The battle went on for 14 hours before the fighting stopped, and although the Jewish quarter had held, no-one doubted that unless help was forthcoming the quarter would soon capitulate. But that evening, a party of 35 Palmach fighters managed to break through the siege lines, and they marched into Safed two abreast, singing the Palmach battle song. Food was distributed and their leader, Elad Peled, a 21-year old company commander in the Yiftach Brigade, assumed control of the Jewish quarter. The Jews now had seven light machine guns, one 2in mortar and 270 rifles, plus a few sub-machine guns and grenades. They had very little ammunition.

The Arab advantage in weapons, men and defe[n]sive positions was so great that Adib Shishakli, th[e] Arab commander in Safed, decided to wait for th[e] Arab invasion to begin, rather than waste his r[e]sources in forcing the Jews to surrender. Accordin[g]ly, he limited his offensive to employing small grou[ps] to attack and harass the Jewish quarter.

The defence of the Jewish quarter was conce[n]trated around three strongpoints – the Central Hot[el,] the Commercial Centre and the Technical Scho[ol] (which was destroyed early on by mortar fire). T[he] fate of the second of these, the Commercial Centr[e,] hung in the balance when Arabs built a tunnel from [a] house in their own quarter, 100yds away, with th[e] intention of coming up under the centre and dyn[a]miting it. The Jews remained blissfully unaware [of] this development, even when the tunnel had pro[g]ressed to within a few yards of the target. Then a 25[lb] shell fired from the Jewish mortar fell exactly ov[er] the extreme end of the tunnel, caving in the eart[h.] The Arabs, convinced that they had been disc[o]vered, abandoned work, yet the Jews did not find t[he] tunnel until after the fighting was over. The shell w[as] a fluke and had not been aimed anywhere near [to] where it landed.

The most exposed spots in the long, twisted Jewi[sh] defence line were a series of buildings standing [at] the edge of the Jewish quarter, less than 20yds fro[m] the Arab positions, which the defenders calle[d] 'Stalingrad', in honour of another, more famou[s] siege. Each evening, groups of Jews would take u[p] positions in the ruined buildings of 'Stalingrad', [to] stand guard all night, tensely waiting for an Ara[b] attack. Arabs and Jews would trade insults, and whe[n] the insults ran out they would fire at each other. An[d] so it went on, night after night.

As the siege worsened, there was talk of evacua[t]ing the women and children from the Jewish quart[er] to Haifa, which by then was already in Jewish hand[s.] The Palmach commander refused to consider th[is] idea, and long after the war he explained why:

'It was very hard to say 'no'...I knew I was assuming a tremendous responsibility, but I had to take the chance. If the children had left, morale would have dropped and we needed every bit of confidence we could call upon. Besides, I could not spare the troops to escort the hundreds of children and their mothers through the perilous mountains into safety.'

Another difficult decision concerned the isolated settlements lying between Safed and the neighbouring Arab states. Many were under attack and requesting urgent help, but Allon knew that control of Safed was vital to the outcome of the war. He therefore concentrated his limited forces on the town, trusting, rightly as it turned out, that this move would divert Arab troops away from the settlements. The attack on Safed was ordered to begin.

The 3rd Battalion of the Yiftach Brigade, commanded by Moshe Kelman, assembled on Mount Canaan, ready to move into the Jewish quarter, from where they would attack the Arab sections of Safed. First, however, they had to break the siege, and this they decided to do at the weakest spot in the Arab siege line – the villages of Ein Zeitun and Biriya on the town's northern flank.

At 0300 on 30 April, two companies of the 3rd Battalion set off to attack Ein Zeitun. The attack began with a barrage from 3in mortars, and from the 'Davidka', the Palmach's primitive home-made mortar. At the same time, all Jewish positions in Safed opened fire in an attempt to pin down Arabs and prevent them from aiding the village. Fierce fighting followed for about 90 minutes, after which the Arabs broke and ran. By dawn, the Palmach held Ein Zeitun.

Meanwhile, a party guarding the route connecting the village with Biriya came under heavy fire from the Teggart fortress on Mount Canaan. The leader of the seven men realised that his choices were to retreat, thus abandoning his mission of protecting the Jewish flank, or to take cover in Arab-held Biriya. The only way they could achieve the latter was by fighting their way in and taking over the village. So they did. With both villages in Jewish hands, morale in the Jewish quarter soared. But the Palmach was still a long way from securing Safed.

Kelman's plan first called for the capture of the Citadel, which overlooked the whole town

As elements of the 3rd Battalion began entering the Jewish quarter, Kelman assumed command of all forces in the town and began making frantic preparations to capture the Arab sections. By 6 May the forces were ready to move. Kelman's plan first called for the capture of the Citadel, which overlooked the whole town. This position would provide a springboard from which to attack the town. A platoon was detailed to capture the Training School on the eastern side of the Citadel, after which a second platoon would advance up to the Citadel and charge the well-defended outpost at its southeast corner. A third platoon was detailed to launch a diversionary attack on the Teggart fortress.

Beginning at 0100 hours on 6 May, the attack went well at first and the Training School was taken without difficulty. But when the second platoon charged up the hill it was pinned down by heavy fire from several different strongpoints. The battle lasted three hours and, although reinforcements were sent up, the force was unable to reach the Citadel. At dawn, Kelman gave the order to retreat: the Palmach

Above: A member of the Palmach maintains a watch on Arab positions with a Bren gun. During the period of the British Mandate the men covered their faces to avoid recognition and capture by the authorities. Above left: Armed with 9mm Sten sub-machine guns, members of the Haganah, now the official army of the new Jewish state, seek vantage points in a bombed building in Haifa. The port is facing attack from the Lebanese and Syrians to the north, and the Iraqis and the Transjordan Arab Legion to the east. Far left: Haganah members in largely British apparel search an Arab village. Left: Jeep-mounted Palmach strike units regroup in a captured Arab village during the drive to oust Egyptian forces from the Negev desert.

In 1947 the British, wearying of their conflict with the Jews in the Mandate of Palestine, handed the problem to the United Nations. The General Assembly voted to partition the territory into two separate states – one Jewish and one Arab – and immediately hostilities broke out as the Arabs attempted to lay the Jewish minority under siege. In the period preceding the British withdrawal in May 1948 the Jews suffered considerable losses but managed to keep vital routes open and convoys moving.

Following the British withdrawal on 15 May, 37,000 troops from Egypt, Transjordan, Syria, Iraq and Lebanon invaded the new state, intent on destroying it. The IDF succeeded in containing the invasion forces, albeit at a heavy cost. The Egyptians advanced up the coast, but were prevented from reaching Tel Aviv, and while the Transjordanian Arab Legion captured the eastern half of Jerusalem, it was unable to gain the rest of the city.

After a four-week ceasefire imposed by the UN that ended on 9 July 1948, lower and western Galilee was cleared of Arab forces, the Arab towns of Ramle and Lydda were captured, and the siege of western Jerusalem lifted. The final stage of the war began in mid-October 1948 and lasted until January 1949. Israeli forces captured Beersheba and cleared the Negev desert, reaching the outskirts of El Arish and Rafah in northern Sinai. Eilat was occupied and the Israelis finally gained the upper Galilee.

On 7 January 1949 the Egyptians were the first to agree to calls for an armistice. A ceasefire came into effect, and in July Syria signed the final armistice agreement. The armistice lines remained Israel's borders until the Six Day War of 1967.

had lost two dead and 18 wounded. The men in the Training School, under pressure to withdraw, mined the building before they left and it exploded as the first Arabs entered it. However, the failure of the attack was a serious blow to morale in the beleaguered Jewish quarter.

Allon left his command post and entered Safed to determine what had gone wrong. Then, together with Kelman, a second plan was drawn up. It was decided to attack three strongpoints simultaneously – the Citadel, the Shalva house and the municipal police station – since in the first attack the Citadel had been effectively supported from these positions. With the expected Arab invasion of Palestine only days away, this time there was no margin for error.

A heavy rain was falling, and it slowed progress almost as much as the intense machine-gun fire

At 2100 hours on 9 May, a bag of explosives was fired from a Davidka, signalling the start of the attack. It landed on a bunker protecting the Citadel – sheer luck, since there was no way of guaranteeing the Davidka's range. The company assigned the capture of the Citadel assembled in Safed's Central Hotel, after which the platoon that was to spearhead the assault began to climb the steep slope up to the top of the hill. A heavy rain was falling, and it slowed progress almost as much as the intense machine-gun fire coming from the Arabs entrenched in the Citadel. Close to the top of the hill, the platoon was pinned down. Unable to advance any further, they found some cover and began to return the fire. The commander was wounded, and his deputy, taking over command, ordered the platoon's Piat anti-tank weapon to be carried up to them. The darkness prevented an accurate estimation of the range, but the third shell scored a direct hit. The platoon then climbed up the slope to take the Citadel without any further difficulty. They were joined by the remainder of the company and began to fortify their position and prepare for a counter-attack.

The Shalva house was defended by about 60 soldiers of the Arab Liberation Army (ALA), mostly Iraqis, armed with three machine guns. The Palmach company assembled in a house 50yds from their objective and opened fire with their Piat and rifles. Under this covering fire, the company demolition team – men whose job it was to place explosives against the walls of the target and blow a hole through which the men could enter – advanced up to the house. The explosives were placed but the rain had made them unserviceable. Avraham Licht, the company commander, decided to ignore this setback and gave the order to charge the position. The ALA soldiers fought off the charge and then ran, but not before Licht had been killed in the assault.

The battle for the police station was the most vicious. The position was defended by some 100 ALA troops and was surrounded by barbed wire and a stone wall. The spearhead of the attacking company – a rifle squad, a demolition squad and the Piat squad – advanced as far as the stone wall, 30yds from the actual building. The demolition squad then attempted to use Bangalore torpedoes on the barbed wire, but the torpedoes did not work and the two men who had been trying to operate them were wounded by fire coming from the building. The attackers were forced to climb over the barbed wire and were now faced with the task of blowing a hole in one of the walls. The defenders were using every weapon they

had, and each member of the demolition team wa[s] wounded as he tried to place his explosives. But th[e] explosives were soaked and did not work. The Pia[t] operators and most of the rifle squad had also bee[n] wounded, possibly by a stray Davidka shell which had exploded in mid-air above them. The platoon commander and the platoon medic could only con[-] tinue to give them support by firing at the building[.] Soldiers who had been held in reserve were rushe[d] to take the place of the wounded, and Kelman, at hi[s] command post in the Central Hotel, sent out men t[o] look for more explosives.

At 0230 hours on 10 May, a new demolition ma[...]

arrived, bringing with him a different type of explosive which, it was hoped, would not be affected by the rain. It worked, and the riflemen managed to blow two huge, irregular holes in the side of the building. The 15 fighters of the attacking company who remained unwounded were then split into two sections. The first entered the building led by Yitzchak Hochman, the company commander, while the other section covered their entry, returning the defenders' fire.

Inside the police station, the fighting was carried out floor by floor. It was fierce and often confused. Hochman was killed in the fighting for control of the first floor and it was his deputy who secured the level, but at heavy cost. Many were wounded, for their equipment was not suited to fighting within an enclosed, fortified concrete building. The new company commander called for, and very quickly received, reinforcements in order to clear the building's upper floors.

It took a desperate effort on the part of the company commander – who, as was the practice in the Palmach, led from the front – to lead his exhausted, wounded men up to the other floors for further vicious fighting. It was not until dawn that the third – and final – floor was cleared.

At one stage, the Palmach brought up a home-made flame-thrower and attempted to burn the door down

Arab resistance, however, had not ended. Shutting a heavy iron door behind them, 19 men barricaded themselves on the roof of the building and fired on the Palmach troops milling about outside. During the day various attempts were made to dislodge them, but all failed. At one stage, the Palmach brought up a home-made flame-thrower and attempted to burn the door down, but all this succeeded in doing was nearly setting the building on fire and seriously wounding Elad Peled, who happened to be passing the door as the flame-thrower was operated. (He survived to become a major-general in the Israeli Defence Forces.)

After several hours, eight wounded Arabs came down off the roof and surrendered. The remaining 11 continued to hold out until dusk, when they were able to climb off and slip away into the darkness.

Safed's principal strongpoints were now in Palmach hands. Their capture not only ended the siege once and for all, but also gave the Jews excellent positions from which to advance into the Arab sections. It was necessary to attack quickly, before the Arabs could recover from the shock of their defeat. Company by company, the 3rd Battalion moved towards their objectives, only to find that they did not have to use their weapons. During the night the Arab citizens of Safed, together with the surviving ALA soldiers, had fled the town. Arab Safed was virtually deserted. (The Arabs manning the Teggart fortress had also fled, although the Palmach would not be aware of this for some time.)

The battle for Safed has been called one of the most fierce of the 1948 Arab-Israeli War. When the Arab armies invaded the new state of Israel a few days after the battle for the town, they had a hard time of it. The question is not whether Safed contributed to the Israeli defence, but rather what would have happened had Safed been in Arab hands at the time of the invasion. From that point of view, and given Safed's strategic value, the battle for the town may be said to have been one of the most decisive and most worthwhile of the war.

THE AUTHOR Jeffrey Abrams is a freelance editor and writer based in Israel. He is a former associate editor of the IDF Journal, the quarterly English-language review of the Israeli Defence Forces.

Palmach soldier, Palestine 1948

The illegality of the Haganah obliged it to apply to various sources for the Palmach's equipment. This soldier wears British khaki drill trousers, an American shirt, a civilian woollen cap and a red-and-white shemagh worn as a scarf. He has a US Army cartridge belt, canteen and cover. His weapon is a Model 24 rifle, the Czech version of the German 7.92mm Kar 98k carbine.

Top left: A roof-top machine-gun post is set up to harass Arab troop convoys. Above left: Palmach members pose with citizens of Safed after the siege. Left: The slopes on which Safed was built dominated the main communications links of central Galilee.

Operating in the hazardous mountain terrain of the Radfan, the para-trained gunners of the Royal Horse Artillery used their expertise to direct crushing firepower into the very midst of Quteibi positions

FIRE
IN THE HILLS

'HE WILL NEVER SURRENDER in battle and will endure shocking wounds, crawling away to die on his own rather than seek aid from his enemy. His territory is such that a few men can hold up a battalion. He is fanatically independent, a local saying being, "Every tribesman thinks himself a Sultan." Unless a settlement is made that will allow him his independence the Quteibi will take to the hills and it will all start again.'

So wrote Harry Cockerill, a former officer in the SAS, describing 'the enemy' in South Arabia. These few words neatly encapsulate the nature of a conflict into which, at its height, thousands of British and Arab troops were committed in an attempt to pacify a tiny tract of land inhabited by some of the finest guerrilla fighters in the world – the Quteibi tribesmen of the Radfan mountains. Supporting these troops were the guns of the Royal Horse Artillery (RHA), and one regiment in particular, the 7th Parachute Light Regiment.

The origins of the Radfan campaign lay in two incidents that took place in the southwest corner of the Arabian peninsula during the early 1960s. The first of these was a coup d'état in Yemen in September 1962. This saw the overthrow of the traditional ruler, the Imam. His regime was supplanted by a left-wing government heavily reinforced by troops and arms supplied by President Gamal Abdel Nasser, Egypt's head of state. Repeated calls from the Yemen to the citizens of the Federation of South Arabia (FSA) for their support in the forthcoming battle against British authority led to the second incident, in December 1963. Following months of unrest and outbreaks of violence, a grenade was thrown at the British High Commissioner, Sir Kennedy Trevaskis, as he prepared to fly from Khormaksar airport, in Aden, to London for a high-level conference. Ironically, this conference had been designed to defuse tension in the FSA. Following the grenade incident, a state of emergency was declared.

Although there was a substantial British garrison in Aden, it was not strong enough to quell the 'Wolves of the Radfan', as the Quteibi were known. The rebels were armed by the Yemenis, and had become incensed when British troops had tried to prevent them from plundering cargo and passenger traffic on the Dhala road between Aden and Sana'a, the capital of Yemen. The Quteibi and the British had been eyeing each other suspiciously since the 1880s, and now the Yemenis had the perfect tool with which to contest control of the border area between Yemen and the FSA, which was some 60 miles up-country from Aden in the heart of the Radfan mountains.

After July 1963 the Quteibi became particularly restless, and reinforcements were called for. Elements of 16 Parachute Brigade had been taking turns to rotate through garrison duties in Bahrain, where they would be on hand to support Kuwait should Iraq repeat her threatening behaviour of 1962. The brigade's gunners were 7 RHA, a battery of which supported each of the three parachute battalions wherever they went. The 1st Battalion, The Para-

Far left: Assisted by Bombardier Ulyett (left of picture), Captain Hugh Colley calls down fire on the Quteibi positions. Below left: Gunner Gerald Walsh feeds another round into the breech of his pack howitzer. Bottom: Gunner Bill Taylor carefully plots his targets. Below: The gunners of G (Mercer's Troop) Battery let fly another 33lb of high explosive.

7 RHA

The 7th Regiment, Royal Horse Artillery (7 RHA), was formed in 1961. Up until this time, there had been four RHA regiments in the Royal Artillery, together with 33 Parachute Light Regiment which served in the airborne role with 16 Parachute Brigade. When it was decided that there should be only three RHA regiments, the 33 Parachute Light Regiment was disbanded, along with 2 and 4 RHA. This left the new unit, 7 RHA, to take over the airborne role. The nine senior batteries of the Royal Horse Artillery were then formed into the three regiments – the 1st, 3rd and 7th. By reorganising the RHA in this manner, it was intended to create an elite cadre through which Royal Artillery officers and men would pass in order to experience the ultimate in professional gunnery. After 16 Parachute Brigade was disbanded in 1977, 7 RHA served in West Germany for four years, with the role of rapid deployment being handed over to the 4th Field Regiment, Royal Artillery. In late 1983, 5 Airborne Brigade was formed and 7 RHA returned to the airborne role as the brigade's artillery regiment. Each battery permanently supports a battalion in the brigade: F (Sphinx) Battery supports 2 Para; G (Mercer's Troop) Battery supports the 2nd Battalion, 2nd King Edward's Own Goorkhas; and I (Bull's Troop) Battery supports 3 Para. Since returning to the airborne role, 7 RHA has re-equipped with the Royal Ordnance 105mm light gun.
Above: The Royal Horse Artillery insignia.

chute Regiment (1 Para), was supported by F (Sphinx) Battery; 2 Para by G (Mercer's Troop) Battery; and 3 Para by I (Bull's Troop) Battery. In July 1963, 3 Para and Bull's Troop were in residence at Hamala Camp, Bahrain. These two units visited the Radfan in July 1963. Only one troop from the RHA battery, under Captain Hugh Colley, was deployed with 3 Para, since the conflict was seen primarily as an infantryman's war. The 3 Para group had been on permanent alert during its deployment to Bahrain, and when the order came to be ready to move in three hours, the battery commander, Major David Drew, was unruffled. A battery hockey match had just begun; it would last about 90 minutes. Since the battery was ready to move at 90 minutes' notice, Drew ordered the game to continue. Hugh Colley's troop of three 105mm pack howitzers left on time.

Very little action was seen, however, during this pre-emergency tour, and the battalion and artillery troop returned to Aden just over one month later. In early 1964 the situation boiled over again. This time, the men of 3 Para group, commanded by Lieutenant-Colonel Anthony Farrar-Hockley, were up to their necks in it.

Once the guns were in position, sangars of rock had to be built to protect them from sniper fire

The first opportunity for action fell to B Company, 3 Para, under the command of 45 Commando, Royal Marines. On 3 May 1964, following a cancelled parachute drop into the mountains in support of 45 Commando, the company fought its way from the local base at Habilayn up Wadi Taym to take a hill codenamed 'Cap Badge', overlooking the village of El Naqil. It was the non-airborne J Battery, 3 RHA, that supported the marines and paras during this battle, and it encountered problems that would plague 7 RHA's batteries in the future. The most crucial of these was the difficulty of moving guns around in the harsh, hot terrain while being fired upon by Quteibi snipers.

Two weeks after B Company's baptism of fire, the remainder of 3 Para, together with the guns of Bull's Troop, were in position at Habilayn, 10 miles south-west of El Naqil and christened 'Pegasus Village' by the paras. Farrar-Hockley's job was to seize and secure the Bakri Ridge, one of the dominant features protecting the Quteibi stronghold of Wadi Dhubsan. At first, this operation was carried out by an ad hoc formation known as 'Radforce'. This was later replaced by a more orthodox formation, 39 Infantry Brigade.

Following their arrival at Habilayn, Bull's Troop's guns had been sited to support operations on the entire Bakri Ridge area. When Farrar-Hockley received his orders to attack the Bakri Ridge, however, he ordered his guns moved to the friendly end of the ridge, on a feature overlooking Wadi Taym.

Moving the guns was a tricky, sometimes miserable business. Wherever possible, the guns were towed along roads by Land Rovers. Where this proved impossible, the guns, ammunition and supplies had to be manhandled or carried by helicopter as underslung loads. Neither of these techniques was straightforward. In the hot, thin air found near the tops of the 4000-5000ft peaks, helicopters sometimes experienced difficulties lifting light loads, let alone guns, gun crews and ammunition. On more than one occasion, a move had to be delayed because the Belvedere and Wessex helicopters were simply

unable to lift a gun the final 500m up to its firir position. The gun would have to be left with a sma team who would haul it up the hill.

Work such as this was difficult at the best of time and could be back-breaking in temperatures th reached 120 degrees Fahrenheit in the shade. Th task of moving weapons was occasionally too toug even for man-handling, and the guns had to b dismantled and the components physically carrie up to the position. A gun barrel was a four-man loa its cradle a two-man load, and ammunition weighe 33lb per round; the gunner would carry two rounds a time from the dropping-off point to the gun positio Gunner (now Battery Sergeant-Major) Alan Be remembers vividly the grinding work of supportir 3 Para on Bakri Ridge. The guns had to be hauled u the last 500m to their dispersed fire positions c adjoining crests, and everything else had to b physically carried up the steep slopes. Once th

guns were in position, sangars of rock had to be built to protect them from sniper fire. For the gunners themselves, two or three-man sangars were constructed in defensive positions around the perimeter of the gun site.

The men lived off standard British Army 10-man

Below: The men of A Troop, F (Sphinx) Battery, seen here shortly before setting off on an operation with 45 Commando. They are, from left to right: Captain R.C. Letchworth, Gunners Scott, Nicklin and Tucker, and Lieutenant-Bombardier Ridge. Main picture: Gunners go into action in the baking heat of the Arabian sun. Maps of the Radfan area were notoriously unreliable, hampering proper ranging, but the gunners of 7 RHA still managed to get most of their shots dead on target.

7th Parachute Light Regiment, RHA
The Radfan, May 1964

Wadi Boran

Monk's Field

Danaba

— Cap Badge

Danaba basin

Wadi Taym

El Naqil

Pegasus

Shab Tem

Bakri Ridge

Hajib

Habilayn

Wadi Rabwa

Lethoom

Wadi Bigair

Arzuqm

Wadi Misrah

Thumier

Matil Fawq

Bayn al Gidr

Hadija Mogga

La Adhab

Wadi Dhubsan

Mas Hagar

Jebel Huriyah

Key
◼ Major 7 RHA bases
➡ Attack on Wadi Dhubsan

Wadi Bulbah

Wadi Tramare

Wadi Dhuraa

Y E M E N

Radfan

Ajra

Dhala

Taiz

FEDERATION OF SOUTH ARABIA

Mocha

RED SEA

Lahej

Aden

GULF OF ADEN

37

TURNING FULL CIRCLE

The experience of the 2nd Battalion, The Parachute Regiment (2 Para), during the Falklands conflict of 1982 provided an interesting and somewhat ironic postscript to the Radfan campaign. When 16 Parachute Brigade was disbanded in 1977, the 7th Parachute Light Regiment, Royal Horse Artillery (7 RHA), reverted to a ground role in West Germany with the British Army of the Rhine. The 105mm pack howitzers, being obsolescent by this time, were sold off. The regiment was then re-equipped with 155mm FH70 guns.

In late 1983, 7 RHA returned to the airborne role and once again became a para gunner unit. Two years later, the commanding officer of the regiment, Lieutenant-Colonel Richards, visited Bull's Troop during a two-week field-firing exercise in the Falklands. Outside the Headquarters of Land Forces Falkland Islands were displayed two 105mm pack howitzers that had been captured from the Argentinian 4th Artillery Regiment during the battle for Goose Green. For Lieutenant-Colonel Richards, something about these guns seemed familiar...he took a closer look.

When Richards recognised the ordnance marks as British, his curiosity was further aroused. He took a pen knife and scraped some of the camouflage paint off the gun shield. Just where he had expected to find it, Richards saw the old Bull's Troop insignia. During the Falklands campaign, 2 Para had been shelled by the very same guns that had supported 3 Para in the Radfan!

compo ration packs, while the water, carried up to the position at first light every day before it got too hot, came in jerricans. Although the long tours of Bahrain and Kuwait, together with the strenuous training undertaken by all parachutists, had acclimatised the gunners to the insufferable heat and had kept them fit, it was still a hard life. When not in action, the gun crews loafed around in their sangars under shelters made from blankets and ponchos, while the gun sentry hugged the shade under the camouflage nets that draped over the guns. Boredom was the worst part of it, that and the incessant waiting for fire orders from the Forward Observation Officers (FOOs) attached to 3 Para and the other army units, and the Naval Gunfire Observers (NGOs) attached to 45 Commando. With so many of the Quteibi otherwise engaged, there were few attempts to snipe at the gunners during this period, and many of the men were disappointed not to have set eyes on the enemy.

Things soon livened up for the gunners when 3 Para began clearing the ridge on 18 May. The battle went on for eight days, with the gunners of Bull's Troop landing their high-explosive rounds perilously close to the embattled infantrymen up front. Despite the tough conditions, the gunners played their part to perfection. It is easy to take accurate artillery fire for granted, without ever considering the very real difficulties and dangers that the gun-

ners face. On Bakri Ridge, the gunners of Bull's Troop performed superbly. Most of the credit for taking the ridge has gone to 3 Para, but Farrar-Hockley was not slow in praising his own gunners for their part in breaking the tribesmen's resistance. Clearing guerrillas out of their own backyard – and the Quteibi were some of the finest guerrillas the British Army has ever encountered – was a monumental task, and 7 RHA deserves enormous credit for its precision shooting.

The FOOs were no less deserving of praise. Moving up with the forward troops, they pinpointed enemy strongpoints and then called in high-explosive fire – relying on the blast and the shrapnel, screaming wickedly around the arid rocks, to kill or at least terrify the enemy. It worked beautifully. The final chapter in 3 Para's tour of duty was the capture of the Wadi Dhubsan itself. This was achieved simultaneously with the capture of the rebel redoubt at Jebel Huriyah by the 1st Battalion, Royal Anglians, and the 2nd Battalion, Federal Regular Army (an Arab force containing a number of British officers). In both cases, the Quteibi made the mistake of trying to hold their ground. The Wadi Dhubsan assault involved a 3000ft descent off the Bakri Ridge, some of it by abseiling, and was a classic example of a battalion attack that was supported by RAF Hunters and the guns of Bull's Troop. With X Company, 45 Commando, under command, Farrar-Hockley led the batta-

Below left: Gunners Saunders, Murphy and Bowen man one of the 105mm pack howitzers from Bull's Troop. Below: A Wessex helicopter airlifts one of F Battery's guns from Monk's Field up onto the Bakri Ridge. Bottom: Men from the Royal Army Service Corps attach a supply load of artillery shells to the lifting winch of a Belvedere helicopter. Below, far left: The Yemen hills provide the backdrop for the gunners of G (Mercer's Troop) Battery, stationed at Ajra, in the north of the Radfan. This picture was taken during the Christmas of 1965, and shows Gunner Savage receiving a 'trim' from Gunner Hartland.

lion off the ridge and, for the first time in their lives, the Quteibi experienced a full-scale British Army assault. By nightfall the Quteibi had folded. The same occurred at Jebel Huriyah, where J Battery's 105mm pack howitzers, together with the 5.5in field guns of 170 Medium Battery, RA, helped pound the tribesmen into submission.

The first round from each 'stonk' had to be right on target – first time every time

Bull's Troop, I Battery, returned to Bahrain shortly after the Radfan campaign ended. The next unit from 7 RHA to deploy to South Arabia was Mercer's Troop, G Battery. This unit withdrew in early 1965. It was followed by the whole of F (Sphinx) Battery, supporting its normal accompanying troops from 1 Para. The battery, commanded by Major Richard Ohlenschlager, RHA, arrived at Khormaksar Airport in Aden on 19 May 1965. The guns and vehicles arrived on 1 June, and the battery split up into its two constituent troops, each comprising three guns. A Troop went up-country to Habilayn, and spent most of June and July helping 45 Commando keep a watchful eye on the Quteibi. B Troop went even further up-country to Monk's Field, near Dhala. Here they supported 45 Commando, then the 2nd Battalion, Coldstream Guards, and the 4th Battalion, Royal Anglians. June and July were difficult months for B Troop. The men came under incessant sniper fire, but responded with accurate artillery shelling and succeeded in killing some of the enemy. The snipers were persistent, however, and sometimes managed to creep as close as 50m before being driven off by smallarms fire from the gunners and the guards.

A Troop had a more leisurely tour of duty. The unit moved to the outposts of Ajra (north of Dhala and within sight of the Yemeni border) and Hayaz for short deployments before being pulled back to carry out internal security duties in 'Little Aden'. Following this, A Troop returned to the United

Kingdom in October 1965. B Troop then took over responsibility for Ajra and Hayaz, and it was at Hayaz that the battery fought its most significant action of the tour. When the Quteibi in this area became restless, 45 Commando was sent in to pacify them. This involved B Troop in a series of lightning moves by helicopter and road that resulted in a significant victory for the battery and the commandos. Enemy casualties were almost certainly very high, although exact figures were hard to ascertain as the Quteibi always recovered their dead and wounded from the battlefield.

Returning to the comparative calm of Dhala, B Troop played an important role in fostering local public relations on the evening of 4 September, during a night-firing exercise. The Emir's palace had come under attack from an indeterminate number of rebels armed with bazookas. The gunners responded instantly, shifting their fire onto the rebels and repelling the attack with heavy enemy casualties. Rebel activity continued to remain at a high level, and Dhala, Ajra and Hayaz were not easy postings for the battery. However, the SAS had set up covert Observation Posts (OPs) in the mountains along the Dhala Road. Together with normal infantry patrols, the SAS were thus able to call fire down on unsuspecting Quteibi rebels and insurgents of the Front for the Liberation of Occupied South Yemen and the National Liberation Front. The quality of the battery's gunnery was extremely high, as the guns could not afford the luxury of ranging shots against such fleeting targets. The first round from each 'stonk' had to be right on target – first time every time.

F (Sphinx) Battery was relieved by B Battery, 1 RHA, on 12 October and returned to its base at Lille Barracks, Aldershot, where it rejoined 16 Parachute Brigade. The gunners of G (Mercer's Troop) Battery did two short, separate tours of the Radfan during 1965-66. The first was from 2 April to 10 June 1965, when they were relieved by F Battery, and the second was from 23 November 1965 to 4 January 1966. Their deployment was similar to that of F

Battery. The enemy was highly active during the first tour, frequently attacking the various army bases in numbers of up to 60 or more. Later on, as the Habilayn garrison (D Company, 1 Para) and other Federal Regular Army (FRA) and British Army troops struck back, these attacks slackened off. Nevertheless, the risk to the British troops was real and immediate, and this danger was increased when the rebels resorted to a determined campaign of road mining.

In early May 1965 the first relief took place, with the original gun crews returning to Bahrain. Captain Morgan of C Troop, on his last day at Ad Dimnah, found himself acting as FOO for X Company, 45 Commando, and had his prayers answered when his patrol spotted 10 of the enemy out in the open, some distance from cover. He called down fire from the guns of 19 Light Regiment, RA, up at Dhala. Morgan then called in an air strike just for good measure. He arrived back in Bahrain feeling that honour was better satisfied.

On 26 May Mercer's Troop joined forces with the SAS up in the Wadi Mishwarrah, during operation 'Mish-Mish'. While a patrol from 22 SAS moved down the wadi, southwest of Dhala, 4 FRA established a base for two of the guns at Wa'alan. After 10 helicopter flights, the troop (less one gun that was left at Ad

Below: Their 105mm pack howitzers hitched up to Land Rovers, the men of F Troop, I (Bull's Troop) Battery, prepare to move out and occupy forward positions. Each man carries his own supply of water, heavily salted to replace the salt lost in the fierce heat. Bottom: the parachute-trained gunners of Mercer's Troop keep a close watch for adventurous guerillas during a move into the mountainous terrain of the Radfan.

Dimnah) moved in. But all the preparation was in vain; when the SAS patrol flushed out the enemy, they were just out of range of the guns.

Ad Dimnah was abandoned on 2 June, after it was realised that the target was too tempting for the enemy. Besides, the road was still being mined with depressing regularity. For the first time since the regiment had deployed to the Radfan, three guns occupied the same site at Habilayn. C and D Companies of 1 Para, having seen Mercer's Troop in action, waited eagerly for their 'own' gunners from F (Sphinx) Battery to relieve them on 9 June.

Mercer's Troop returned to the Radfan in November 1965. D Troop went straight up to Ajra and there completed the work begun by F Battery, turning the base into a neat three-gun position. Although artillery and mortar fire could be heard coming from the other side of the border, none came the way of Mercer's Troop and only one attack took place – a half-hearted affair by about 20 rebels on 7 December. It lasted 15 minutes, during which time the troop fired 36 rounds at ranges of 800m, rising slowly as the rebels fled. C Troop relieved D Troop on 18 December, and Christmas was celebrated with a carol service in one of the gun pits, the gun's trails forming the base for the altar table. With no further action, the troop returned to Bahrain on 4 January 1966.

During their service in South Arabia, the gunners of 7 RHA had shown themselves to be a highly professional organisation even by the Royal Artillery's own high standards. Although there were other regiments of the Royal Artillery present in the area during this time, 7 RHA's lonely months up at Habilayn, Ad Dimnah, Ajra and Monk's Field were vital for the support of the troops – mainly paras, marines and FRA – who were trying to quell the insurgents. Being parachute gunners, the men of 7 RHA knew what to expect when deployed to remote areas without support themselves, but responsible for supporting the infantry. Like the commando gunners who also took part in these campaigns, the para gunners had to combine the traditional solid reliability demanded by infantry commanders of their supporting arms, with the versatility of the infantry themselves. It was not easy, but parachute gunners do not go for 'cushy' postings – they go where the work is.

THE AUTHOR Gregor Ferguson was a former editor of *Defence Africa and the Middle East* and has contributed to several other publications. His most recent work is a short history of The Parachute Regiment with which he served in the 10th (Volunteer) Battalion.

In May 1981, a potentially explosive hostage situation was defused by GEO, Spain's formidable anti-terrorist unit

AT 0900 HOURS on Saturday, 23 May, 1981, the ornamental gates of the Banco Centrale, an imposing granite building in Barcelona's main square, the Plaza de Cataluna, swung open for business. It was another perfect summer's day, already warm, and Barcelona was bustling with tourists. The nearby Ramblas was already open for business, the market fresh with the scent of flowers and the song of cage birds for sale. The bank soon filled with customers and, counting the staff, its cool, marble halls held some 200 people.

No-one took any notice of the men who drifted in and spread themselves out at strategic points through the main hall. They looked like any other customers – until, that is, they pulled dark balaclava helmets over their heads, pointed guns at the staff and customers and ordered them to put their hands in the air and keep quiet.

They still had nearly 200 hostages in their power – more than enough for their purposes

The orders were given by one man, who, throughout the operation was addressed by his men as 'Numero Uno' – Number One. Just to prove how serious they were, they shot one of the bank clerks, Ricardo Martinez Calafell, in the leg. However, another of the clerks managed to press an alarm connected to the local police station. A further employee, arriving late for work, saw his colleagues with their hands in the air, turned and fled, also raising the alarm. Within minutes, police cars, their sirens echoing round the fashionable shops and banks that line the vast plaza, surrounded the Banco Centrale. The first impression of the police was that they were dealing with a highly professional gang of bank robbers. But it soon became evident that this was no ordinary robbery.

The first account of what was happening inside the bank came from the wounded Calafell who was allowed to leave the building, followed soon afterwards by 21 of the hostages who were suffering from shock. Their release did not matter to the gunmen; they still had nearly 200 hostages in their power – more than enough for their purposes. The story gradually emerged of some 20 men, armed with automatic weapons, who had wired up the building with explosives and were threatening to kill their hostages if certain demands were not met. These demands were discovered in a note left in a nearby telephone box. Their purpose, said the note, was to 'finish Red terrorism'. The gunmen wanted the release from prison of Lieutenant-Colonel Antonio Tejero, Major-General Luis Torres Roja, Colonel Jose Ignacio San Martin and Lieutenant-Colonel Pedro Mas Oliver. And they wanted two aircraft

ALL THE KING'S MEN

41

put at their disposal, one at Barcelona airport for their own use and the other at Madrid to fly the released officers to Argentina.

It was now that the affair became political dynamite, for it was Tejero who, on 23 February of that year, had led some 300 of the Guardia Civil in storming the Cortes, the Spanish Parliament. The world, through the television cameras in the Cortes building, had seen the pudgy, balding, moustachioed Tejero waving his pistol at the members of parliament. By his actions, Tejero had represented the army's discontent with King Juan Carlos' determination to turn Spain into a 20th century democracy following the death of the old dictator, General Franco. For a time, it was touch and go – Tejero was merely the figurehead for a variety of plots within the senior ranks of the army. One of the generals involved was Milans del Busch, commander of the Valencia military region, who, as Tejoro stormed the Cortes, had declared a state of emergency and put his tanks on the streets. An army coup had seemed very near.

The 60 men drew their equipment, prepacked for a siege operation, and set out for Barcelona

However, the King had behaved with exemplary courage and coolness. He ordered del Busch to recall his troops to their barracks, and began telephoning the regional commanders to demand their loyalty to him and to the country. One by one, they gave that support. But the danger remained – as Tejero's occupation of the Cortes continued through the night, 13 jeep-loads of military police from the crack Brunete armoured division joined the mutineers. Meanwhile, the King marshalled his forces in preparation to storm the building and gave orders for GEO, the Special Operations Group, to lead the assault. It would have been the anti-terrorist group's first major operation. However, by noon the next day, with the GEO units moving into position, Tejero realised the game was up and surrendered. Curiously, he has since become something of a cult hero in Spain, even standing for Parliament from his prison cell.

It was against this background of attempted coup and rumoured coup that the drama of the occupation of the Banco Centrale developed three months later.

Thirty police cars and vans with some 300 armed police laid siege to the bank, taking cover behind newspaper kiosks and in the entrance to an underground station. Several ambulances waited ominously in nearby streets. After an inconclusive exchange of shots, a parley was arranged over a telephone link with Numero Uno, resulting in 30 hostages being exchanged for food.

It was now that Prime Minister Calvo Sotelo consulted his crisis co-ordinating committee and ordered GEO to the scene. Alerted at their closely guarded barracks at Guadalajara, 60 men – half of the unit's total strength – were ordered to fly the 460km east to Barcelona. The move left the barracks virtually empty, since the rest of the unit, in their customary

'sticks' of five men, were operating against the ETA terrorists in the Basque country to the north. Led by their commanding officer, a major whose name cannot be revealed, the 60 men drew their equipment, pre-packed for a siege operation, and set out for Barcelona. By now the situation had taken a grim turn. The gunmen had given the government 24 hours to accept their demands – at the end of that time, they would kill 10 hostages and then one every hour until, finally, they would blow up the bank and everyone in it. The authorities had no alternative but to believe that they were serious. However, given the political climate, the government could not surrender to their demands. GEO arrived and immediately began its preparations to assault the bank after calling for the plans of the building.

Throughout the night, GEO troopers wearing yellow oil-skins lifted man-hole covers in the streets surrounding the square and went down into the sewers, taking explosives and sand-bags in an attempt to mole their way into the building. But it had been designed to withstand the explosives and thermic lances of professional bank robbers – there was no way in through the sewers. It was thus decided to launch the assault through as many entrances as possible, thereby opening up a number of escape routes for the hostages. Frame charges, similar to those used by the SAS in the assault on the Iranian embassy in London the previous year, were to be used to blow in doors and window frames.

The television crews and radio reporters who had been conscientiously recording the progress of the

Using sophisticated radio equipment to keep in constant contact with their commanding officer, two members of GEO align the sights of their G3 SG/1 sniper's rifles.

siege and disseminating it in great detail – meaning that the gunmen, as well as the rest of Spain, knew precisely what was going on – were ordered to close down. Instead, the incongruous strains of *Auld Lang Syne* echoed round the square, mixing with the bells from the nearby cathedral that were summoning the faithful to prayers on Sunday morning.

The preparations for the assault went on all through that hot Sunday. GEO's sophisticated electronic probes, designed to pinpoint the position of people inside buildings, were as unable as the 'moles' to find a way through the bank's solid granite walls. Fortunately, however, the released hostages had given a fairly accurate picture of what was going on inside. Most of the hostages were being held in the main hall where they had been formed into a

Below, far left: Assault training. Speed is of the essence, and on the order to move in GEO operatives will storm their assigned objective.

human barricade, but others, mainly bank employees, had been taken to the magnificent chandeliered banqueting hall. There were, said the hostages, about 25 armed men inside the building and they had placed explosives round the windows. It was obvious that GEO was going to have a hard time, especially if, as it was then thought, the gunmen were mutinous members of the Guardia Civil – all highly trained men. Nevertheless, the assault went in at 2020, just 35 hours after the gunmen had first walked into the bank. GEO 'sticks' slipped onto the terraces of the bank from a nearby building and burst their way in through side entrances and windows.

Firing broke out at once, as the gunmen shattered windows and shot at GEO troopers assaulting the front door. Hysterical hostages, meanwhile, waved white handkerchiefs from other windows and screamed at the troops not to shoot.

The assault on the front door came to a halt when the hostages who had been formed into a human barricade by the gunmen rushed it and burst out into the square. GEO troopers hustled them away to safety while marksmen gave them covering fire, their bullets knocking chips off the granite building. Six of the hostages lay prone in the open until, using a break in shooting, they scrambled to their feet and ran for cover. Meanwhile, the troopers who had got into the bank worked their way through it, room by room, using the 'killing house' technique they had been taught by the SAS. Operating in two-man teams, kicking open doors and going through to take out the gunmen, each man covered his colleague, with both taking instantaneous decisions on whether to shoot or hold their fire.

'I had one gun against my face and another in my neck as I was pushed out'

It says much for GEO's training, expertise and discipline that only one man was killed during the assault, and he was one of the gunmen. None of the hostages was hurt. Most of the hostages – it was estimated that there were about 70 left in the bank – now began escaping through the open front doors, hands on head, running bent double along the pavement and into the safety of a side-street. They were then searched by police in case they were escaping gunmen, before being helped into police vans. Bomb disposal teams followed the assault sticks through the bank to render harmless the explosives believed to have been planted in the building. The last bastion of the gunmen to fall to GEO was the banqueting room. As the troops burst in, the hostages shouted out for them not to shoot because they were employees of the bank. GEO was taking no chances. Again, following the example of the SAS at the Iranian embassy, they treated the hostages as potential gunmen. One of the freed employees later complained, 'I had one gun against my face and another in my neck as I was pushed out.' GEO had good cause to be wary. Of the 25 gunmen supposedly inside the bank, they had killed one and arrested ten. Although the hostages may have over-estimated the number of gunmen inside the building, it seems possible that some escaped. When the siege was over, hostages spoke of gunmen discarding their hoods, abandoning their weapons and mingling with their prisoners in order to escape. One even insisted on exchanging shirts with one of the hostages.

Another mystery was the paucity and age of the weapons left behind in the bank. The official tally was

one sub-machine gun, eight pistols, two revolvers and six flick knives – and all the guns were over 30 years old. What may be significant is that the number of firearms captured adds up to 11, precisely the number of gunmen accounted for. While the siege was still going on, Lieutenant-Colonel Tejero, who the gunmen described as 'this hero', disowned their attempt to free him. In his customary flamboyant style, he issued a statement through his lawyer saying he had no intention of fleeing from his pending court-martial. The gunmen were certainly not Basques, because one of the hostages had asked his captors whether they were members of ETA, the Basque terrorist organisation. 'ETA', growled the gunman, 'We hate ETA.' So who were they?

At first, the government announced that the gunmen were common criminals or anarchists but, in the face of national incredulity, revised its statement to say that the gunmen were mercenaries – some with criminal records for bank robbery – who had been offered six million pesetas (some £32,000) each at a secret meeting at Perpignan, in southern France, to undertake the operation to free Tejero and his fellow conspirators. A more sophisticated explanation was that they were hired not with any real hope of freeing Tejero, but as part of a series of events designed to shake confidence in the government. This view was given added force by the discovery that Jose Maria Cuevas, the gunman killed by GEO, had rented an apartment along the route that King Juan Carlos was due to take through Barcelona on the Armed Forces Day Parade a week after the siege. When searching the apartment, the police found the beginnings of a tunnel only 30yds from a spot the King would pass. No

Above: Terrified civilians throw themselves to the ground as GEO operatives prepare to storm the bank. Covered by a colleague (right), a member of GEO motions the hostages to safety (inset right). Above centre: Armed and ready for action. Above, far right: Practising marksmanship on the firing range.

ETA

One of the most violent of European terrorist movements is Euskadi Ta Askatasuna (ETA) – which translates as Basque Homeland and Liberty. In 1953, a group of students founded EGIN (With You) in support of the Basque Nationalist Party. Six years later, radicals within the organisation formed ETA with the intention of undertaking terrorist action. Their operations began in 1961, being carried out by ETA-Militar (ETA-M), with a strength of 100 terrorists. Initially, ETA benefited from the fact that the Basque country straddles the Franco-Spanish border. With the French turning a blind eye to ETA activities, the terrorists were able to set up bases and safe houses in France. However, this situation has changed dramatically and several known terrorists – including 'Txomin' Domingo Iturbe Abasolo, the reputed head of ETA – have been extradited from France. Since 1970, ETA has married the old Basque goal of an independent state to the Marxist-Leninist ideology of socialist revolution. By undermining the stability of the Spanish liberal democracy, ETA hopes to provoke a military takeover, which, theoretically, would result in an uprising by the Spanish people.

ETA pursues its objectives in three directions: by exacting a 'revolutionary tax' from rich industrialists; attacking government projects in the Basque area; and waging a war of assassination against the police and armed forces. Its most spectacular operation was the killing of Admiral Carrero Blanco in December 1973. A 100kg bomb, hidden in a tunnel, blew Blanco's car over a four-storey building. Despite the splits within the organisation and the impact of substantial concessions from the government, ETA-M continues its campaign of violence. Able to draw on substantial nationalist support, it seems unlikely to give up.

45

explosives were found, but the assumption was that it was the beginning of a bomb attempt on the King's life. He showed admirable courage in going through with his parade. It did not pass unnoticed that the Plaza de Cataluna, with the flags of Spain and Catalonia proudly waving side by side, bore a very different aspect to that of a week before. GEO had returned to the Plaza, but to protect the King not to storm the bank. To onlookers it seemed that both the King and his protectors had been tested and had been found equal to that test.

The siege at the Banco Centrale had been GEO's first public engagement, made in the full glare of publicity and, despite the questions that still remain unanswered, it had performed as it was designed to perform – efficiently and responsibly, with the minimum of fuss. Since its baptism of fire, GEO has gone back to doing its duty the way it likes best, undercover in small operations. However, it is now so active in the fight against terrorism that it can no longer escape publicity. It was GEO that rescued Dr Julio Iglesias, father of the pop singer, in a spectacular raid on a village house in Zaragoza, 250km west of Barcelona, where he was being held for a million-pound ransom by ETA terrorists. In a similar operation, GEO freed the Basque industrialist, Señor Juan Pedro Guzman, from an ETA 'people's prison' and captured three terrorists without even firing a shot.

Other GEO successes have been more bloody. In March 1984 they ambushed five leading ETA terrorists as they paddled ashore in a rubber dinghy at the harbour of Pasajes, north of San Sebastian. The terrorists had embarked from a French beach a few miles north in order to carry out a car-bombing and a kidnapping. Four of them died in a hail of bullets while the fifth was wounded and captured. Follow-

Below: Back at barracks, GEO operatives continue their training on the firing range, using MP5s and a belt-fed machine gun. Used by the Spanish security forces as a spearhead in the struggle against ETA, GEO normally operates in northern Spain and, although the details of its activities remain shrouded in secrecy, one officer estimated that the unit is called upon at least four times each month. Armed with the Heckler and Koch range of weapons, drawn from the barrack's logistics warehouses, GEO continues the fight against terrorism.

up operations led to the discovery of four 'safe' houses and the capture of a large cache of explosives, arms, money and documents. Never before had so many ETA terrorists died in a single clash with the authorities.

There is no doubt that, despite its short history, GEO has come of age and can take its place among the best of the world's anti-terrorist units.

THE AUTHOR Christopher Dobson is a lecturer in terrorism at the Bramshill Police Staff College and writes extensively on terrorism and military affairs. He is co-author of the book *War Without End*.

Streaming from their transports, the strike force of Israeli paras tore into the airport at Entebbe to do battle with a group of terrorists

SHORTLY BEFORE dawn on 3 July 1976, specialist units of the Israeli Defence Forces (IDF) loaded their equipment and drove to a nearby airbase, where ground-crews stood ready to lash their vehicles into the bellies of four Hercules transports. Elsewhere on the field, the crew of a Boeing 707, fitted out as a mobile hospital, made their final checks. By early afternoon, the transports were airborne and heading for Ophira on the southern tip of the Sinai peninsula; the Boeing began the first leg of a journey that would take it to the airport at Nairobi in Kenya. Operation Thunderball, the plan to rescue 105 Jewish hostages held by terrorists at Entebbe, Uganda's international airport, was underway.

The crisis faced by the Israelis had begun at 1230 hours on 27 June, when four terrorists, two members of the German Baader-Meinhof gang and two members of the Popular Front for the Liberation of Palestine, hijacked Air France Flight 139, with 12 crew and 246 passengers on board, as it flew from Tel

THUNDERBALL AT ENTEBBE

PREPARATIONS

In the immediate aftermath of the hijacking of Flight 139 on 27 June 1976, the government of Israel began formulating its response to the terrorists' demands. Once it became clear that the lives of the Jewish passengers were in danger, the authorities started to consider a military response. However, before an appropriate course of action could be thrashed out, the Israelis had to build up a detailed picture of the situation at Entebbe.

The first break came on 30 June, when a batch of non-Jewish hostages were released by the hijackers. After arriving at Orly airport near Paris, all were interviewed by undercover agents with regard to the number of terrorists, their weapons, routine and clothing, and the degree of Ugandan involvement.

As agents gathered intelligence, a high-ranking team, consisting of politicians and military chiefs, discussed the range of possible military options. Three main rescue strategies were covered: a paradrop into Lake Victoria followed up by a waterborne assault, a direct assault from Kenya; and an airborne landing on Entebbe. Although the third response was considered most feasible, no firm decision was made until further information was available.

It was discovered that an Israeli firm had constructed several installations at the airport and agents were sent out to 'borrow' the blueprints. Other men paid several visits to travel agencies and airline offices to enquire about the regular flight schedules over East Africa. As the Israeli Air Force had helped to train the Ugandans, several members of the missions were contacted to provide intelligence on the lay-out of Entebbe, the location of other airfields, and the strength of Amin's air force. By early July, the Israelis had a detailed picture of the airport and a workable rescue plan. However, the final decision to launch the rescue bid was not taken until 3 July.

Above left and left: Israeli paras in training.

47

Entebbe
Israeli paras, July 1976

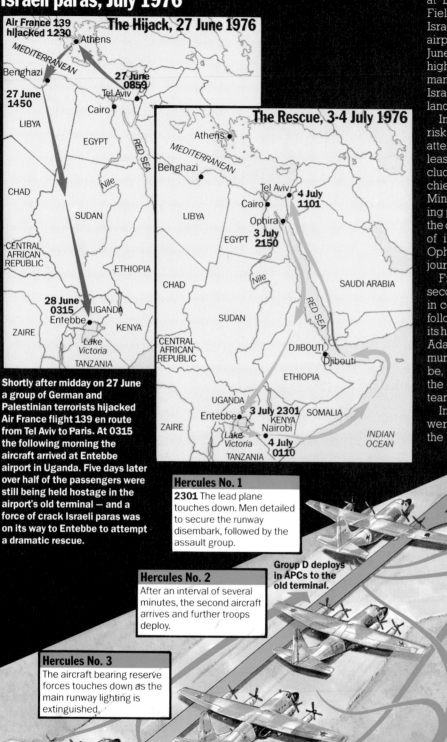

The Hijack, 27 June 1976

Air France 139 hijacked 1230
Athens
MEDITERRANEAN
Benghazi
27 June 0859
Tel Aviv
27 June 1450
Cairo
LIBYA
EGYPT
RED SEA
Nile
CHAD
SUDAN
CENTRAL AFRICAN REPUBLIC
ETHIOPIA
28 June 0315
Entebbe
UGANDA
KENYA
ZAIRE
Lake Victoria
TANZANIA

The Rescue, 3-4 July 1976

Athens
MEDITERRANEAN
Benghazi
Tel Aviv
4 July 1101
Cairo
Ophira
3 July 2150
LIBYA
EGYPT
CHAD
Nile
SUDAN
SAUDI ARABIA
RED SEA
CENTRAL AFRICAN REPUBLIC
DJIBOUTI
Djibouti
ETHIOPIA
UGANDA
Entebbe
3 July 2301
KENYA
Nairobi
SOMALIA
ZAIRE
Lake Victoria
4 July 0110
TANZANIA
INDIAN OCEAN

Shortly after midday on 27 June a group of German and Palestinian terrorists hijacked Air France flight 139 en route from Tel Aviv to Paris. At 0315 the following morning the aircraft arrived at Entebbe airport in Uganda. Five days later over half of the passengers were still being held hostage in the airport's old terminal — and a force of crack Israeli paras was on its way to Entebbe to attempt a dramatic rescue.

Hercules No. 1
2301 The lead plane touches down. Men detailed to secure the runway disembark, followed by the assault group.

Hercules No. 2
After an interval of several minutes, the second aircraft arrives and further troops deploy.

Hercules No. 3
The aircraft bearing reserve forces touches down as the main runway lighting is extinguished.

Hercules No. 4
2308 The final Hercules is on the ground. Further reserve forces disembark and the aircraft taxies towards the old terminal to pick up rescued hostages.

Group D deploys in APCs to the old terminal.

Group B secures the main runway and takes the new terminal and control tower.

Route of Hercules Nos. 1 - 3
Route of Hercules No. 4

Command Group and move out from the fir aircraft and assume c

Group A drives down t taxiway and assaults old terminal.

Group C, the reserve force, moves down to the old terminal on foot to assist with the evacuation.

Key
Operation Thunderball
Operation Thunderball: routes of Hercules transpo
Operation Thunderball: routes of assault groups
Air France flight 139

Aviv for Paris via Athens. After the takeover, the passenger jet was diverted to B█████i in Libya, where it was refuelled, and then f███ █h, landing at Entebbe at 0315 on the 28th. Uganda █ ██er, Field-Marshal Idi Amin Dada, was no friend of the Israelis and welcomed the terrorists, who used the airport's old terminal to hold the hostages. On 29 June, the hijackers, organised and supported by a highly-developed international terror network, demanded the release of 53 of their comrades held in Israel, France, West Germany, Kenya and Switzerland.

Initially, the Israeli government was unwilling to risk the lives of the non-Jewish hostages in a rescue attempt, but when the other passengers were released, senior politicians and military leaders, including Lieutenant-General Mordechai Gur, the chief-of-staff, Prime Minister Yitzhak Rabin and Minister of Defence Shimon Peres, accepted a daring plan proposed by Major-General Dan Shomron, the director of infantry and paratroopers. After a day of intensive preparation, the assault teams left Ophira airbase on 3 July; ahead lay a 3000-mile journey to Entebbe.

Fifteen minutes after the last aircraft left Ophira, a second Boeing was on its way south from an airbase in central Israel. It would also land at Ophira then follow the transports – three hours behind to allow for its higher speed. On board were Major-General Kuti Adam, another senior officer, and a team of communications officers. Their job was to circle Entebbe, providing a vital radio link between the men on the ground, the planners in Israel and the medical team at Nairobi airport.

In the cockpits of the four transport planes which were now flying low over the Gulf of Suez, beneath the reach of hostile radar surveillance, the pilots

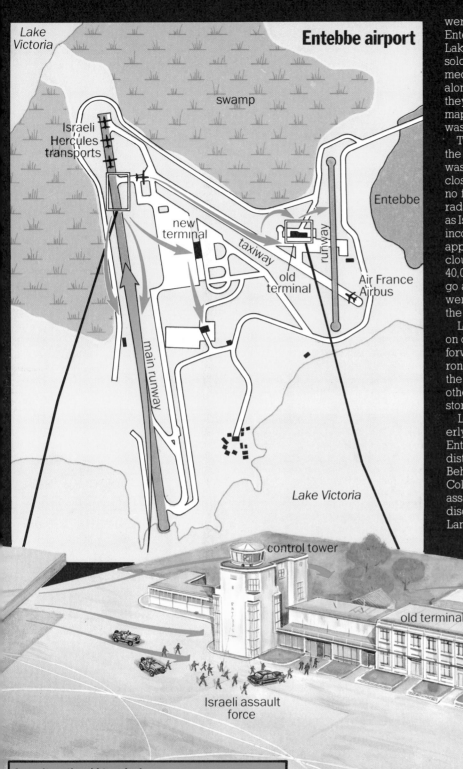

Lake Victoria

Entebbe airport

swamp

Israeli Hercules transports

new terminal

taxiway

old terminal

runway

Entebbe

Air France Airbus

main runway

Lake Victoria

control tower

old terminal

Ugandan Air Force MiG-17s

Israeli assault force

were studying a batch of aerial photographs of Entebbe airport taken from Kenyan airspace over Lake Victoria. In the bellies of the aircraft, the soldiers of the assault teams, and the doctors and medics who were to land with them, sprawled alongside their vehicles catching whatever sleep they could. Some of the officers were studying their maps and orders again, making sure that everything was committed to memory.

Turning westwards, the four Hercules headed into the African continent over Ethiopia. The weather was stormy, forcing the pilots to divert northwards close to the Sudanese frontier. However, there were no fears of detection: it was doubtful that any alert radar operators would be able to identify the planes as Israeli, and the storm would wreak havoc with any incoming signals on their screens. Later, on the approach to Lake Victoria, the aircraft hit storm clouds towering in a solid mass from ground level to 40,000ft. There was no time to go round, and no way to go above – they just ploughed through. Conditions were so bad that the cockpit windows were blue with the flashes of static electricity.

Lieutenant-Colonel 'S' held the lead plane straight on course; his cargo of 86 officers and men, and the forward command post of Major-General Dan Shomron, with their vehicles and equipment, had to be on the ground according to a precise timetable. The other pilots had no choice but to circle inside the storm for a few extra minutes.

Lieutenant-Colonel 'S' kept the aircraft on a southerly course, then banked sharply to line up on Entebbe's main runway from the southwest. In the distance he could see that the runway lights were on. Behind him, in the cargo compartment, Lieutenant-Colonel Jonathan 'Yoni' Netanyahu's men, the first assault wave, were piling into a black Mercedes, disguised to resemble Amin's personal car, and two Land Rovers. The car engines were already running,

Assault on the old terminal

2301 The assault force drives to the old terminal. When challenged the paras in the Mercedes leading car open fire. The assault goes in: para teams enter the building and engage terrorists and Ugandan soldiers.
2304 The terminal is secured. The fourth Hercules taxies in and the freed hostages are taken on board.
2352 The Hercules bearing the hostages is airborne and the withdrawal of the remaining Israeli forces commences after several Ugandan MiG-17s have been disabled.

4X-FBB

Left: Field-Marshal Idi Amin Dada, Uganda's power-hungry dictator, gave his full support to the hijackers and his intransigence forced the Israeli government to use force to secure the release of the hostages. Ironically, Amin had close links with the Israelis who had trained elements of his armed forces and helped in the modernisation of Entebbe's facilities in the years before the crisis. On ceremonial occasions, as here, he took great pride in wearing the much-coveted Israeli para wings above all other decorations. Above and above right: Speed and deception were the keys to the success of Operation Thunderball. Heavily armed jeeps and a Mercedes, disguised as Amin's personal limousine, were used to fool the Ugandan airport guards. Below: Doctor Jossi Faktor, head of the operation's medical team.

The Mercedes and its escorts moved down the connecting road to the airport's old terminal building as fast as they could, consistent with the appearance of a senior officer's entourage. On the approaches to the tarmac apron in front of the building, two Ugandan sentries faced the oncoming vehicles, aimed their carbines, and shouted an order to stop. There was no choice, and no time to argue. The first shots from the Mercedes were from silenced pistols. One Ugandan fell and the other ran in the direction of the old control tower. The Ugandan on the ground was groping for his carbine. A paratrooper responded with a burst. Muki, Netanyahu's second-in-command, and his team jumped from the car and ran the last 40yds to the walkway in front of the building. The first entrance had been blocked off; without a second's pause, the paratroopers raced on to the second door.

After a searching debate with Netanyahu, Muki had decided to break a cardinal rule of the IDF. Junior officers usually lead the first wave of an assault, but Muki felt it important to be up front, in case there was need to make quick decisions about changes in

MEDICAL AID

The officer in charge of the medical support team, Doctor Jossi Faktor, recalls his part in the Entebbe operation:

'In addition to our usual medical supplies, we carried [aboard our Hercules transport] lots of 'space blankets' (aluminium sheets used for burn wounds) and large old-fashioned milk pails. Both were useful: the sheets were used to cover the hostages and Air France crew who had insufficient clothing, and the pails were used as giant sick bags.

'Being the last Hercules to land at Entebbe and the first to leave, we spent less than one

and members of the aircrew were standing by to release the restraining cables.

At 2301 hours, only 30 seconds behind the pre-planned schedule, Lieutenant-Colonel 'S' brought the aircraft in to touch down at Entebbe in the wake of a scheduled cargo flight that unwittingly covered the first landing. The rear ramp of the Hercules was already open, and the vehicles were on the ground and moving away before the plane rolled to a stop. A handful of paratroopers had already jumped out of the aircraft to place beacons next to the runway lights, in case the control tower shut them down.

As the authorities in Israel considered their response to the hostage crisis, senior military officers began to prepare forces to carry out a rescue mission. In secret, assault teams commanded by Lieutenant-Colonel Jonathan Netanyahu, consisting of the cream of the country's paras and members of the Golani Brigade, gathered at a remote airbase in Israel. Based on available intelligence, a mock-up of Entebbe's old terminal was rapidly constructed and the men practised their assault techniques under the critical gaze of Netanyahu. However, because of the layout of the airport, it became clear that a speedier means of reaching the terminal was imperative. The problem was solved by Netanyahu's second-in-command, who remembered that senior Ugandan officers, and Amin himself, always travelled in black Mercedes limousines. A quick phone call, and a Mercedes was on its way to the airbase.

As the assault groups were being put through their paces, the Israeli Air Force's senior officers picked the pilots and aircrews to fly Netanyahu's men to Entebbe. The distance was no problem, Hercules transports had regularly flown to Uganda. However, there were difficulties in landing at night and the crews were ordered to practise landing in darkness. After several sessions at a remote air strip in Israel, the senior air force officers detailed to oversee the raid training were convinced that their men were equal to the task. Unsure as to the physical state of the hostages the Israelis also made provisions for medical staff to accompany the assault force. In the days before the raid both reservists and current military doctors were placed on alert and a Boeing jet was fitted out as a hospital.

Although given less than a week to prepare, all the elements of the operation were ready for action by 3 July.

plan. Tearing along the walkway, he was fired on by a Ugandan. Muki responded, killing him. A terrorist stepped out of the main door of the old terminal to see what the fuss was, and rapidly returned the way he had come.

Muki then discovered that the magazine of his carbine was empty. The normal procedure would have been to step aside and let someone else take the lead. He decided against this, and groped to change magazines on the run. The young officer behind him, realising what was happening, came up alongside. The two of them, and one other trooper, reached the door together – Amnon, the young lieutenant, on the left, Muki in the centre, and the trooper on the right.

The terrorist who had ventured out was now standing to the left of the door. Amnon fired, followed by Muki. Across the room, a terrorist rose to his feet and fired at the hostages sprawled around him, most of whom had been trying to get some sleep. Muki took care of him with two shots. Over to the right, another member of the hijack team managed to loose off a burst at the intruders, but his bullets went high, hitting a window and showering glass into the room. The trooper aimed and fired. Meanwhile Amnon identified a female terrorist to the left of the doorway and fired.

In the background, a bullhorn was booming in Hebrew and English, 'This is the IDF. Stay down.' From a nearby mattress, a young man launched himself at the trio in the doorway and was cut down by a carbine burst. The man was a bewildered hostage. Muki's troopers fanned out through the room and into the corridor to the washroom beyond – but all resistance was over.

The second assault team had meanwhile raced through another doorway into a hall where the off-duty terrorists spent their spare time. Two men in civilian clothes walked calmly towards them. Assuming that they were hostages, the soldiers held their fire. Suddenly, one of the men raised his hand and threw a grenade. The troopers dropped to the ground. A machine-gun burst eliminated their adversaries and the grenade exploded harmlessly.

Netanyahu's third team from the Land Rovers moved in to silence any opposition from the Ugandan soldiers stationed near the windows on the floor above. On the way up the stairs, they met two soldiers, one of whom opened fire. The troopers killed them.

While his men circulated through the hall, calming the shocked hostages and tending the wounded, Muki was called out to the tarmac. There he found a doctor kneeling over his commanding officer.

hour on the ground. We spent the short flight to Nairobi evaluating and stabilising the condition of the wounded soldiers and hostages. However, our desperate attempts to resuscitate "Yoni" were to no avail.

'At Nairobi, we were faced with the difficult decision to leave some of the more seriously wounded behind as they required immediate hospitalisation. After refuelling, we took off on the final leg of the journey home, which was medically uneventful. The hostages were shocked and excited; the soldiers emotionally exhausted. Yet only a few managed to doze off after 36 hours without sleep. The rest is now history.'

Netanyahu had remained outside the building to supervise all three assault teams and a bullet from the top of the old control tower had hit him in the back. While the troopers silenced the fire from above, he was dragged into the shelter of an overhanging wall by the walkway.

The assault on the old terminal was completed within three minutes after the lead plane had landed. Now, in rapid succession, its three companions came in to touch down at Entebbe. By 2308 hours, all of Thunderball Force was on the ground. The runway lights shut down as the third plane came in to land, but it didn't matter – the Israeli beacons did the job well enough.

With clockwork precision, armoured personnel carriers roared off the ramp of the second transport to take up positions to the front and rear of the old terminal, while infantrymen from the first and third planes ran to secure all access roads to the airport and to take over the new terminal and control tower. The tower was vital for the safe evacuation of the

4X-FBQ

Above: After their triumphant return to Israel on 4 July, the assault teams received their nation's grateful thanks from Major-General Dan Shomron, the director of paratroopers and infantry (standing, third from left), and the Minister of Defence, Shimon Peres (standing, far left). Left: Scenes of unrestrained jubilation marked the arrival of the freed hostages at Lod airport. The celebrations were marred only by the sad loss of Lieutenant-Colonel Jonathan 'Yoni' Netanyahu (far left), cut down during the fighting at Entebbe's old terminal. Below: One of the Israeli Air Force's Hercules transports taxis to a halt on a remote airfield.

hostages and their rescuers. In a brief clash at the new terminal, Sergeant Hershko Surin fell wounded. The fourth plane taxied to a holding position near the old terminal ready to take on the hostages. The engines were left running. A team of air force technicians was already hard at work off-loading heavy fuel pumps to transfer Idi Amin's precious aviation fuel into the thirsty tanks of the lead transport – a process that would take well over an hour.

Meanwhile, as planned, the Medical Corps' Boeing had landed at Nairobi, at 2205 hours. General Benny Peled, the GOC Air Force, was able to radio Lieutenant-Colonel 'S' that it was possible to refuel in Kenya's capital. Unable to raise Shomron on the operational radio, and uncomfortable with the situation on the ground – the Ugandans were firing tracer at random, while the aircraft with engines running were vulnerable at the fuel tanks – he decided to make use of the Nairobi option.

Muki radioed Shomron to report that the building and surroundings were secure, and to inform him that Netanyahu had been hit. Although they were ahead of schedule, there was no point in waiting around (possibly allowing the Ugandans the time to bring up reinforcements). The fourth Hercules was ordered to move closer to the old terminal. Muki's men and the other soldiers around the building formed two lines from the doorway to the ramp of the plane: no chances would be taken that a bewildered hostage might wander off into the night or blunder into the aircraft's engines.

As the hostages straggled out, the heads of each family were stopped at the ramp and asked to check that all their kin were present. Captain Bacos, the civilian pilot of Flight 139, was quietly requested to perform the same task for his family – the crew of the airliner. Behind them, the old terminal was empty but for the bodies of six terrorists, among them Gabriele Kröcher-Tiedemann and a blond-haired man, Wilfried Böse, both members of the Baader-Meinhof gang. Seven other terrorists, who were at Entebbe to meet the hostages when they first arrived, also died.

It took seven minutes to load the precious cargo of humanity, while a pick-up truck, brought 3000 miles specially for the purpose, ferried out the dead and wounded, including Netanyahu. The paratroopers made a last check of the main building, then signalled the aircrew to close up and go. At 2352 hours, the craft was airborne and on its way to Nairobi. Inside, doctors worked over seven wounded hostages. Two

had died during the rescue and a third, Mrs Dora Bloch, moved to a hospital in Uganda's capital, Kampala, before the raid, was later murdered.

At the other end of the airfield, an infantry team fired machine-gun bursts into seven Ugandan Air Force MiGs. There was no point in tempting Ugandan pilots into pursuit. The paratroopers reloaded their vehicles and equipment. Their job done, they were airborne at 0012 hours on the 4th.

The tired airmen in the cockpit were astonished to see people in the streets below waving and clapping

Thirty minutes after the final departure, the communications Boeing and the first Hercules touched down at Nairobi and taxied to the fuel tanks in a quiet corner of the airport. Sergeant Hershko, who was seriously wounded, was transferred to the hospital Boeing. Two hostages whose wounds needed immediate care in a fully equipped hospital were loaded into a waiting station wagon and taken into Nairobi. At 0204 the remaining passengers and crew of Flight 139 were airborne on the last leg of their long journey home. In Lieutenant-Colonel 'S's' aircraft, the paratroopers were sunk in their own private thoughts. Despite all the efforts of the doctors, Netanyahu was dead. The mission was later renamed Operation Jonathan in his memory.

Early in the morning, the lead Hercules flew low over Eilat, at the southern tip of Israel. The tired airmen in the cockpit were astonished to see people in the streets below waving and clapping. The plane flew on to land at an air force base in central Israel. Here, the hostages were fed and given a chance to shake off the trauma. The wounded were taken off to hospital, and psychologists circulated among the rest, giving help where it was needed.

In a corner of the same airfield, the three combat teams unloaded their vehicles and equipment. They would return to their own bases, hardly aware of the excitement in Israel over their achievement.

It was mid-morning when a Hercules transport of the Israeli Air Force touched down at Lod airport outside Tel Aviv, rolled to a stop and lowered its rear ramp to release its cargo of men, women and children into the outstretched arms of their relatives and friends, watched by a crowd of thousands. The ordeal was over.

By the rescue of the hostages, the Israelis had gone some way to showing the rest of the world that terrorism could be met and defeated by the clinical application of controlled force.

THE AUTHOR Major Louis Williams, an editor with long experience in publications on military subjects, is a senior press officer with the Israeli Defence Force Spokesman's reserve unit. A fuller version of this article appears in the IDF Journal of May 1985.

MOUNTAIN TROOPS

In the years before World War I, only a handful of the German Army's Jäger battalions and the men of the 82nd Infantry Brigade received regular training in mountain and winter warfare. During the early stages of the war, the conditions encountered by the troops on the mountainous Vosges front in France led to the creation of more specialised units.

Based on cadres of men from the Jäger corps, five ski battalions were raised – four in Bavaria and one in Württemberg.

The 1st, 2nd and 3rd (Bavarian) Ski Battalions were formed into the 3rd Jäger Regiment of the army's Alpine Corps and, in the early part of 1915, elements of the regiment saw action in Trentino and Serbia before transferring to the Western Front in 1916 for the Battle of Verdun. Later in the year, the unit was sent to the Carpathians in Central Europe. The Württemberg Mountain Battalion was formed during October 1915 and fought in the Vosges until late 1916, when it was transferred to Romania. Following another spell in the Vosges in early 1917, the battalion moved south in September, to take part in the Battle of Caporetto.

Rommel, who had joined the unit at the outset, was placed in charge of an Abteilung, an ad hoc detachment, for the offensive and played a prominent role in the capture of Mount Matajur. The battalion itself was in the forefront of the battle of Caporetto but lost a small number of men. The Württemberg Battalion was increased to regimental strength before the end of the war.

Above: The Edelweiss badge of Germany's mountain troops.

In 1917 a small detachment of mountain troops, under the young Erwin Rommel, took on the Italians at the Battle of Caporetto

AT 0200 hours on 24 October 1917, along a 24-mile length of the Isonzo front in northeast Italy, 1845 guns of the German and Austro-Hungarian Fourteenth Army unleashed a massive bombardment on the Italian frontline positions on the west bank of the river. The Twelfth Battle of Isonzo, better known as Caporetto, was underway. Five-and-a-half hours later, assault troops poured out of their trenches and began to swarm through the region's valleys and across its mountains, striking out for Udine and the Tagliamento river.

One of the Fourteenth Army's four assault gruppen, commanded by General Freiherr von Stein, was ordered to punch a hole through the Italian defences in the area of Monte Matajur and then link up with a second gruppe under General Alfred Krauss for a follow-on push towards the basin of Caporetto. By the evening of the 24th, von Stein's men had carved out a salient on the east bank of the Insonzo and were moving on Matajur. However, the Italians, initially caught wrong-footed by the onslaught, began to resist strongly and the offensive threatened to stall.

To prevent a stalemate, on the 26th elements of Stein's command carried out a brilliant attack on Monte Matajur. Led by the Germans' crack mountain corps, the gruppe surged forward. In the forefront of the assault was a detachment of the Württemberg Mountain Regiment. One of its junior officers, responsible for taking the position, was later to gain great renown in World War II. His name was Erwin Rommel, and he later recorded his part in the

Below: Shells hit Italian positions near Caporetto during the Austro-German offensive of late 1917. Later, an assault troop (right), including a mountain detachment under Erwin Rommel (centre right), crossed the Isonzo river (above right) to take on the Italians beyond. In difficult country, pack mules (top right) were vital. Far right: A trench mortar in action.

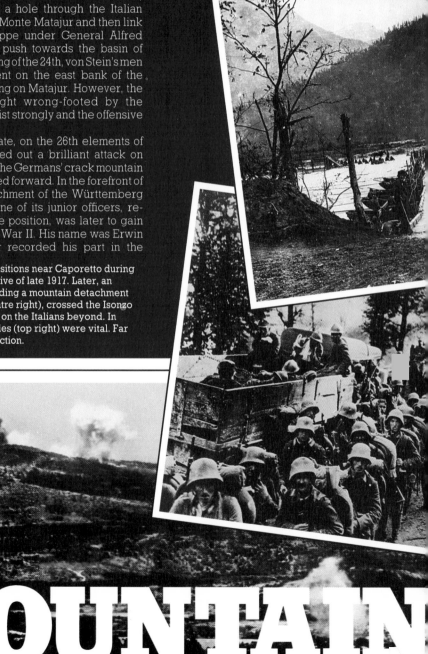

MOUNTAIN

OFFENSIVE

offensive in his book *Infantry Attacks*. Before Rommel's detachment could get to grips with the enemy on Monte Matajur, it had to capture an enemy position on Mrzli peak. Rommel takes up the story:

"By 0830 on 26 October the 2nd Company, having dwindled to a platoon with two light machine guns, captured Hill 1192, a mile-and-a-half west of Avsa. The enemy prevented a further advance. The Italians were in considerable strength half-a-mile northeast of Mrzli peak, and plastered our newly won hilltop with heavy machine-gun fire. A minimum force required to attack the enemy on the southeast slope of the Mrzli, was estimated at two rifle companies and one machine-gun company. In order to assemble these forces quickly, I hurried to the rear down the Matajur road.

Right: Surrounded by scree slopes and precipitous peaks, members of an Austro-German supply column receive rations from a mobile field kitchen. With movement restricted by difficult terrain, regular logistical support was the exception rather than the norm. Bottom left: German infantry, armed with rifles and a Lewis gun, practise assault techniques for the Caporetto offensive. Bottom right: A machine-gun crew fires on an Italian-held position.

Caporetto

On 23 May 1915 Italy joined the Allies by declaring war on Austria-Hungary. In the north, Italy secured the Alpine passes and made a limited advance into the Trentino, but – although the Italians had been promised territory in the south of Austria by the terms of the secret treaty of London – the Italians' main objective was the city of Trieste, at that time a part of Austria-Hungary. Accordingly, the main weight of the Italian offensive was directed towards the east. From June 1915 to September 1917 the Italians mounted eleven offensives against the Austro-Hungarian positions along the Isonzo river. But weak in artillery and short of supplies, the Italian armies had only limited success: they gained some six miles at the expense of 600,000 casualties.

At the end of August 1916 Italy declared war against Germany, Austria-Hungary's ally. The Italians concentrated their divisions well to the south of Caporetto, relying on the natural defences provided by the mountainous terrain in the northern part of the front. On 24 October 1917, the German and Austro-Hungarian armies launched their counter-stroke in the weaker northern sector. The breakthrough at Caporetto, in which the forces under Erwin Rommel played a notable part, quickly became a rout. By early November, the Italians had been driven back to the Piave.

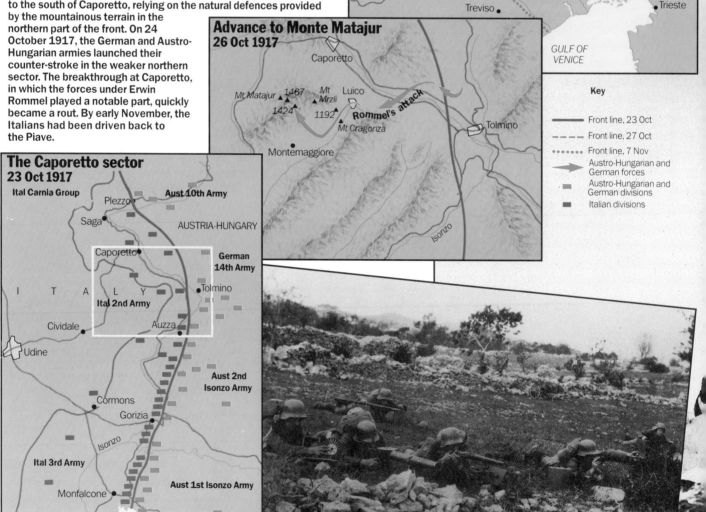

The Italian front 1915 – 1917

Alps
AUSTRIA-HUNGARY
Julian Alps
Bolzano
Cortina
Tolmezzo
TRENTINO
Caporetto
Belluno
Udine
Tagliamento
Trent
Gorizia
Vittorio Veneto
Isonzo
Piave
ITALY
Trieste
Treviso
GULF OF VENICE

Advance to Monte Matajur 26 Oct 1917

Caporetto
Mt Matajur 1467 Mt Mrzli Luico
1424 1192 Rommel's attack
Mt Cragonza Tolmino
Montemaggiore
Isonzo

Key

Front line, 23 Oct
Front line, 27 Oct
Front line, 7 Nov
Austro-Hungarian and German forces
Austro-Hungarian and German divisions
Italian divisions

The Caporetto sector 23 Oct 1917

Ital Carnia Group
Aust 10th Army
Plezzo
Saga
AUSTRIA-HUNGARY
Caporetto
German 14th Army
Tolmino
ITALY
Ital 2nd Army
Cividale
Auzza
Udine
Aust 2nd Isonzo Army
Cormons
Gorizia
Isonzo
Ital 3rd Army
Aust 1st Isonzo Army
Monfalcone

Remaining neutral during the first year of World War I, Italy declared war on Austria-Hungary in May 1915 with the intention of gaining control over the Italian-speaking territories of Trentino in the north and Trieste in the east. The latter region was the focus of Italian ambitions between 1915 and 1916, but a succession of offensives ground to a halt against the mountains of the Julian Alps. Two fruitless years were devoted to gaining a foothold in the alps on the other side of the Isonzo river; an effort that, after the loss of some 600,000 casualties during 11 separate campaigns, came to nought. However, between August and September 1917, the Italians were able to make some headway, forcing the Austrians to call on Germany for military assistance. In response, six divisions, four trained in mountain warfare, were despatched south. The Austrian counter-blast, later known as the Battle of Caporetto, began on 24 October and succeeded in pushing the demoralised Italian Army back to the Piave river, a distance of some 60 miles. Alarmed by the turn of events, France and Britain came to Italy's aid, sending 11 divisions in mid-November.

Although Caporetto was a disaster of the first order for the Italians, it also marked the turning point of the war. Under a new commander, General Armando Diaz, the army recovered and held the Piave line, defeating an Austrian offensive in July 1918.

With growing confidence, Diaz was able to plan a massive counter-attack that was to lead to the collapse of the Austro-Hungarian Army. Begun in late October, the offensive, later known as the Battle of Vittorio Veneto, routed the enemy. At 1500 hours on 4 November an armistice came into effect; the war in Italy was over.

I had to wait until 1000 before I had assembled a force equal to two rifle and one machine-gun company. These groups were composed of men from companies of my detachment. Their approach to Hill 1192 was greatly delayed, because the various units were repeatedly involved in battles with the enemy, who was trying to retreat in a southwesterly direction across the Mount Cragonza–Hill 1192 line.

I felt we were strong enough to engage the Italian garrison on the Mrzli. By means of light signals, we asked for artillery fire on the hostile positions on the southeast slope of the Mrzli peak, with the astounding result that German shells were soon striking there. Then, the lively fire of the machine-gun company from Hill 1192 pinned the hostile garrison down in their positions, while two rifle companies under my leadership came into close combat with the Italians just below the ridge road. We succeeded in turning the hostile west flank. Then we swung in against the flank and rear of the hostile positions. But the enemy hastily withdrew when they saw us attacking in this direction and retired to the east slope of the Mrzli. We took a few dozen prisoners. Since I did not intend to follow the enemy retreating on the east or north slopes of the Mrzli, I broke off the engagement, continued down the ridge road towards the south slopes of the Mrzli and brought up the machine-gun company.

Already, during our attack, we had observed hundreds of Italian soldiers in an extensive bivouac area in the saddle of the Mrzli between its two highest prominences. They were standing about, seemingly irresolute and inactive, and watched our advance as if petrified. They had not expected the Germans from a southerly direction – that is, from the rear. We were only a mile away from this concentration of troops. The Matajur road wound up over the partially wooded south slope of the Mrzli and, on the way west to the Matajur, passed just under the hostile encampment.

The number of the enemy in the saddle on the Mrzli was continually increasing until the Italians must have had two or three battalions massed there. Since they did not come out fighting, I moved nearer along the road, waving a handkerchief, with my detachment echeloned in great depth. The three days of offensive had indicated how we should deal with the new enemy. We approached to within 1100yds and nothing happened. They had no intention of fighting although their position was far from hopeless! Had the Italians committed all their forces, they would have crushed my weak detachment and regained Mount Cragonza. Or they could have re-tired to the Matajur massif almost unseen, under the fire support of a few machine guns. Nothing like that happened. In a dense human mass the hostile forma-tion stood there as though petrified and did not budge. Our waving with handkerchiefs went un-answered.

We drew nearer and moved into a dense high forest 700yds from the enemy and thus out of their line of sight, for they were located about 300ft up the slope. The road bent sharply to the east and we wondered what the enemy up there would do. Perhaps they had decided to fight? If the Italians rushed downhill, we would have had a man-to-man battle in the forest. The enemy was fresh, had tremendous numerical superiority, and moreover enjoyed the advantage of being able to fight down-hill. Under these conditions I considered it a vital necessity to reach the edge of the wood below the hostile camp. But my mountain troopers with the heavy machine guns on their backs were so ex-hausted that I did not expect them to make the steep climb through dense underbrush.

I left the edge of the forest and, walking steadily forward, demanded that the enemy surrender

Therefore, I allowed the detachment to continue marching along the road, while Lieutenant Streicher, Dr Lenz, a few mountain soldiers and I climbed on a broad front, with about 100yds interval between men, and took the shortest route through the forest towards the enemy. Lieutenant Streicher surprised a hostile machine-gun crew and took it prisoner. We reached the edge of the forest unhin-dered. We were still 300yds from the enemy above the Matajur road; it was a huge mass of men. Much shouting and gesticulating was going on. They all had weapons in their hands. Up front, there seemed to be a group of officers. The leading elements were not expected for some time and I estimated them to be 700yds to the east.

With the feeling of being forced to act before the adversary decided to do something, I left the edge of the forest and, walking steadily forward, demanded, by calling and waving my handkerchief, that the enemy surrender and lay down their weapons. The mass of men stared at me and did not move. I was about 100yds from the edge of the woods, and a retreat under enemy fire was impossible. I had the impression that I must not stand still or we were lost.

I came to within 150yds of the enemy. Suddenly, the mass began to move and, in the ensuing panic, swept its resisting officers along downhill. Most of the soldiers threw their weapons away and hun-dreds hurried to me. In an instant, I was surrounded and hoisted on Italian shoulders. 'Eviva Germania!' sounded from 1000 throats. An Italian officer who hesitated to surrender was shot down by his own troops. For the Italians on Mrzli peak the war was over. They shouted with joy.

Now the head of my mountain troops came up along the road from the forest. They moved forward with their habitual easy but powerful mountaineer stride, in spite of the hot sun and their heavy loads. Through an Italian who spoke German, I ordered the prisoners to line up facing the east and below the Matajur road. There were 1500 men of the 1st Regiment of the Salerno Brigade. I did not let my own detachment halt at all, but I did call one officer and three men out of the column. Two mountain riflemen

were assigned to move the Italian regiment across Mount Cragonza to Luico; and the disarming and removal of the 43 Italian officers, separated from their men, was entrusted to Sergeant Goppinger. The Italian officers became pugnacious after seeing the weakness of my detachment and they tried to re-establish control over their men. But it was too late. Goppinger performed his duty conscientiously.

While the disarmed regiment moved down to-wards the valley, my men moved past just below the Italian camp ground. Some captured Italians had told me shortly before that the 2nd Regiment of the Salerno Brigade was on the slopes of the Matajur; it was a very famous Italian regiment which had been repeatedly praised by General Cadorna, the Italian commander, in his order of the day because of outstanding achievements before the enemy. They assured me that this regiment would certainly fire on us and that we would have to be careful.

Their assumption was correct. The head of my detachment had no sooner reached the west slope of the Mrzli than strong machine-gun fire opened up from Hill 1467 (4842ft) and Hill 1424 (4700ft). The hostile machine-gun fire was excellently adjusted on the road and soon swept it clear. Dense bushes below the road protected us from aimed fire. My men were soon under control and I continued the march, not below the Matajur road in the direction of Hill 1467, but in a sharp turn to the southwest. I wanted to cross Hill 1223 (4035ft) at the double and head towards the hairpin turn in the Matajur road just south of Hill 1424. Once there, then the 2nd Regiment of the Salerno Brigade could scarcely escape and would be in a position similar to that of the 1st Regiment 30 minutes before. The only difference would be that a withdrawal to the south across the bare slopes of the Matajur would be prevented by our fire, whereas on Mrzli peak, a covered retreat through the wooded zone had remained open to the Italians.

In order to deceive the enemy, I ordered a few machine guns to fire from the west slopes of the Mrzli. With the rest of the detachment, I reached the turn of the road, 700yds south of Hill 1424, without under-going hostile fire, as the enemy was unable to observe our movement through the thick clumps of bushes. I prepared a surprise attack on the garrison of Hill 1424, which was still firing on the rearward units of my detachment and on our machine guns on the Mrzli. The success on the Mrzli had caused us to forget all our efforts, our fatigue, our sore feet and our shoulders, chafed by heavy burdens.

The captured colonel fumed with rage when he saw that we were only a handful of men

While I was expeditiously carrying out the prepara-tions for the attack, ordering the machine-gun pla-toons in position, and organising assault squads, the order came from the rear: 'Württemberg Mountain Battalion withdraws.' News of the great number of prisoners taken by my detachment, over 3200 men, had reached my commanding officer and given him the impression that the hostile resistance on the Matajur massif was already broken. The order to withdraw resulted in all units of the detachment marching back to Mount Cragonza, except for the 100 riflemen and six heavy machine-gun crews who remained with me. I debated breaking off the en-gagement and returning to Mount Cragonza.

No! The battalion order was given without know-ledge of the situation on the south slopes of the

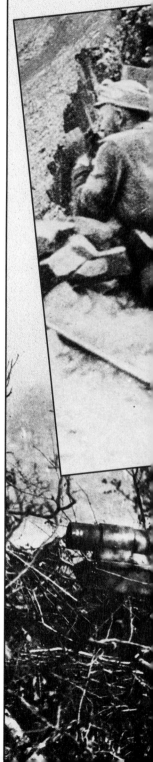

Below: Protected by a sanga and portable armoured shields, a Maxim-gun detachment waits for the observer (left) to point out a new target. Possession of the region's high ground conferred great tactical advantages, allowing troops to bring fire to bear on the enemy in the valleys. Bottom: The crew of a German 10cm gun, relishing the advantage of a commanding position, bombard the retreating Italians.

Matajur. Unfinished business remained. To be sure, I did not figure on further reinforcements in the near future, but the terrain favoured the plan of attack greatly – and every Württemberg trooper was, in my opinion, the equal of 20 Italians. We ventured to attack in spite of our ridiculously small numbers.

Over on Hills 1421 and 1467 the enemy was facing east among large rocks and dived for cover when our unexpected machine-gun fire hit them from the south. The heavy fragmentation up there in the rocks considerably increased the effect of each shot. The hostile reaction was slight. Our machine guns had been emplaced in dense, high bushes, so that the enemy had trouble locating them.

We kept attacking. The heavy machine guns were moved up in echelon. From Hill 1467, a hostile battalion tried to move off to the southwest by way of Scrilo. But the fire of one of our machine guns, delivered at 60yds from the head of the column, forced the battalion to halt. A few minutes later, waving handkerchiefs, we approached the rocky hill 600yds south of Hill 1467. The enemy had ceased firing. Two heavy machine guns in our rear covered our advance. An unnatural silence prevailed. Now and then we saw an Italian slipping down through the rocks. The road itself wound among the rocks and restricted our view of the terrain to a few yards. As we swung around a sharp bend, the

Württemberger, Caporetto 1917

This man's green collar patches, piping and shoulder 'rolls' denote membership of the Württemberg Mountain Battalion, as does the Edelweiss emblem on his bergmütze cap. Armament comprises a 7.92mm M98 carbine.

Right: Dejected Italian prisoners, some of them taken during the campaign, wait to be shipped off to POW camps. The victors (below right) enjoy a hurried meal before continuing their drive to the Piave river, a few dozen miles to the east.

view to the left opened up again. Before us, scarcely 300yds away, stood the 2nd Regiment of the Salerno Brigade. Its men were assembling and laying down their arms. Deeply moved, the regimental commander sat at the roadside, surrounded by his officers, and wept with rage and shame over the insubordination of the soldiers of his once-proud regiment. Quickly, before the Italians saw my small numbers, I separated the 35 officers from the 1200 men so far assembled, and sent the latter down the Matajur road at the double, towards Luico. The captured colonel fumed with rage when he saw that we were only a handful of soldiers.

Without stopping, I continued the attack against the summit of the Matajur. The latter was still a mile away and 700ft above us, and we could see the garrison in position on the rocky summit. They apparently did not intend to follow the example of their comrades on the south slopes of the Matajur who had surrendered and were marching away. Lieutenant Leuze used his few machine guns to give fire support for the attack which we attempted on the shortest route from the south. But the hostile defensive fire was very heavy there and the avenues of approach were so disadvantageous that I preferred to turn to the east on the arched slope, unseen by the enemy, and attack the summit position from Hill 1467. During this movement small squads of Italians, with and without weapons, kept on moving towards the spot where the 2nd Regiment of the Salerno Brigade had laid down its arms.

Before we opened fire, the garrison on the summit gave the sign of surrender

We surprised an entire Italian company on the sharp east ridge of the Matajur, 600yds east of the peak. In total ignorance of events in its rear, it was on the north and was engaged with scout squads of the 12th Division, who were climbing towards the Matajur from Mount Della Colonna. Our sudden appearance on the slope in the rear, with weapons at the ready, forced these men to surrender at once.

While Lieutenant Leuze fired on the garrison of the summit with a few machine guns from a southeasterly direction, I climbed with the other units of my small group in a westerly direction, along the ridge and towards the summit. On a rocky knoll a quarter-of-a-mile east of the peak, other heavy machine guns went into position as fire support for the assault team disposed on the south slope. But before we opened fire, the garrison on the summit gave the sign of surrender. Over 120 more men waited patiently, until we took them prisoner at the ruined building, a border guardhouse, on the summit of the Matajur, Hill 1641 (5414ft). A scout squad of the 23rd Infantry Regiment, consisting of a sergeant and six men, met us during its climb from the north. At 1140 on 26 October, three green flares and a white one announced that the Matajur massif had fallen."

THE AUTHOR Erwin Rommel, undoubtedly one of the finest commanders of World War II, served with the Württemberg Mountain Regiment at Caporetto and wrote of his exploits in his book *Infantry Attacks*.

60

TO THE LAST ROUND

Faced with the crushing might of a German cavalry division at Néry in 1914, the brave gunners of L Battery fought to the last man and the last shell

Below: Royal Horse Artillery gunners prepare their 13-pounder QF (quick-firing) for action during the retreat from Mons in late August 1914. Bottom: RHA crewmen training at the gallop.

NCO, Royal Horse Artillery, Néry 1914

This sergeant wears regulation uniform: khaki serge service dress, breeches with leather pads, puttees wound from knee to ankle, and standard ammunition boots fitted with spurs. Badges, worn on his universal-pattern forage cap and shoulder straps, indicate that he is a member of the horse artillery. Equipment consists of a riding whip, a 1903-pattern bandolier and straps for a waterbottle and haversack.

ROYAL GUNNERS

Since the Napoleonic Wars the artillery of the British Army has been held in the highest regard by friend and foe alike. Within this prestigious arm of service the Royal Horse Artillery (RHA) claims a special elite status – a fact duly recognised in the army's order of precedence, in which the guns of the RHA claim the premier position on the right of the line. The RHA fulfilled a hybrid role by combining the panache of the light cavalryman with the technical competence of the professional artilleryman.

By 1914 the men of the RHA had built up a gallant tradition of self-sacrifice on the battlefield, not only in defence of their guns (regarded by the artilleryman as the embodiment of his unit), but in the selfless provision of fire support for the other arms of service. In 1914 the RHA was organised into six-gun batteries, equipped with quick-firing 13-pounder field guns, each of which was drawn by a team of six horses.

L Battery traced its origins back to India and the 3rd Troop of the Bengal Horse Artillery which was raised in 1809. The troop fought with distinction in many of the conflicts on the Indian sub-continent during the 19th century: in the mountains of Nepal against the Gurkhas; against the Sikhs in the Punjab; and in the Indian Mutiny of 1857 where the unit won its first Victoria Cross. In 1861 the 3rd Troop was absorbed into the Royal Artillery as L Battery, RHA. At the outbreak of war in August 1914, L Battery was assigned to the 1st Cavalry Brigade. After the Battle of Mons on 23 August the British Army was compelled to retire in the face of vastly superior German forces, and the 1st Cavalry Brigade was given the role of protecting the move.

Above: The badge worn by the Royal Horse Artillery in 1914.

AS DUSK fell on the evening of 31 August 1914, the men and horses of L Battery, Royal Horse Artillery, wearily arrived at the little French village of Néry. L Battery was part of the 1st Cavalry Brigade (a flank guard of the British Expeditionary Force), now in full retreat from superior German forces. Earlier in the day, reconnaissance had established that advance elements of the German First Army were fast closing on the British. The 1st Cavalry Brigade had prepared an encampment at Néry and its three regiments, the 2nd Dragoon Guards (Queen's Bays), the 5th Dragoon Guards and the 11th Hussars, were billeted in and around Néry, while L Battery bivouaced in a small field outside of the village.

The following morning revealed a thick mist, reducing visibility to around 150yds. A stand-fast was ordered until 0500 hours, while patrols were sent out to the north and east of Néry. L Battery, comprising six 13-pounder field guns plus limbers and ammunition wagons, was 'stood to': poles were let down and, in the expectation of a long, hot march ahead, the battery horses were watered, one section at a time. First-line ammunition wagons were away, replenishing their stocks and many of the men of the battery were dispersed around the bivouac area. Around 0500 the battery commander, Major Sclater-Booth, walked over to brigade HQ in the centre of the village to receive the day's orders. As he walked through the door, a patrol officer breathlessly informed HQ that he had ridden into a body of German cavalry and had been chased back to Néry. As he was speaking, a high-explosive shell burst over the village and firing began from the commanding heights to the east of Néry.

The battery position was a shambles: broken guns and limbers littered the ground

---

The 1st Cavalry Brigade had been completely surprised: under cover of night and the early morning mist, the German 4th Cavalry Division had stolen a march over the British, so that by early morning a full-scale assault was being launched against Néry. Three German four-gun batteries were in position less than 1000yds from the village; their fire was joined by a machine-gun detachment and smallarms from the dismounted cavalry. In Néry itself pandemonium broke loose: maddened horses galloped up and down the main street, as the troopers and gunners of the brigade tried to organise themselves under a withering hail of fire. Sclater-Booth ran out of HQ, but before he could regain his battery, a shell exploded in front of him, knocking him unconscious. He was discovered after the battle, temporarily blinded.

Meanwhile, L Battery was bearing the brunt of the fire. Encamped on an exposed area on the eastern side of the village, it was overlooked by the German gunners, who regarded the enemy artillery as a prime target. Captain Bradbury was standing with the battery's three other officers when the first shrapnel round burst in the air over them. Several more shells followed; the horses bolted, overturning wagons and carts and destroying one of the guns. Within seconds, two more guns were out of action. The battery position was a shambles: broken guns and limbers littered the ground, along with wounded and dying horses. By the end of the battle 150 horses had been killed.

Bradbury realised the gravity of the situation and, after calling out, 'Come on! Who's for the guns?', he

ran out towards the 13-pounders, followed by t[ ] other officers and about a dozen men. As shrapn[ ] and high explosive rained down on them, the office[ ] and men of L Battery managed to unlimber the thr[ ] serviceable guns and bring up ammunition. Eve[ ] then, the ammunition wagons were some 20 or 30y[ ] behind the guns and each round had to be brought[ ] over this bullet-swept ground. 'F' gun was taken [ ] Captain Bradbury, 'C' gun was under Lieutena[ ] Gifford and 'D' gun was controlled by Lieutenar[ ] Campbell and Munday.

Unfortunately, before 'D' gun could fire a sing[ ] round it was hit by a German shell and silenced. T[ ] survivors – Campbell, Munday and Bombardi[ ] Perrett – ran across to help bring up ammunition to[ ] gun. 'C' gun began firing at the German guns on t[ ] plateau some 800yds away, but after firing only eig[ ] rounds, a shell exploded over the gun's muzzle. T[ ] crew was either killed or wounded, with the exce[ ] tion of Lieutenant Gifford, who carried on sing[ ] handedly until all the ammunition from the limb[ ] had been fired. By this time Gifford had been woun[ ] ed three times and then, as he dragged himself ba[ ] to the shelter of some haystacks in the corner of t[ ] field, he was wounded a fourth time.

However, Captain Bradbury's 'F' gun bore [ ] charmed life. Despite the Germans' best efforts, the[ ] could not silence the gun and it kept up its fire as lo[ ] as there was ammunition. It was clear that it w[ ] having an effect, for the shrapnel bursts were se[ ] erely mauling the German gun crews. All the whil[ ] the gunners were supported by dismounte[ ] cavalrymen, keeping up a solid fire with their Vic[ ] ers machine guns and Lee Enfield rifles.

The rapid fire maintained by 'F' gun took its toll [ ] the crew, as Gunner Darbyshire related:

'As soon as we got "F" gun into action, I jumpe[ ] into the seat and began firing, but so awful was t[ ] concussion of our own explosions and the burs[ ] ing German shells that I could not bear it long[ ] kept it up for about 20 minutes, then my nose an[ ]

Below: Operating in the open countryside of northern France, a horse battery opens fire on the advancing German Army. The 13-pounder, weighing 2236lb, was introduced into service in 1904 and performed well in t[ ] opening stages of World War I; it was later overshadowed by the 18-pounder.

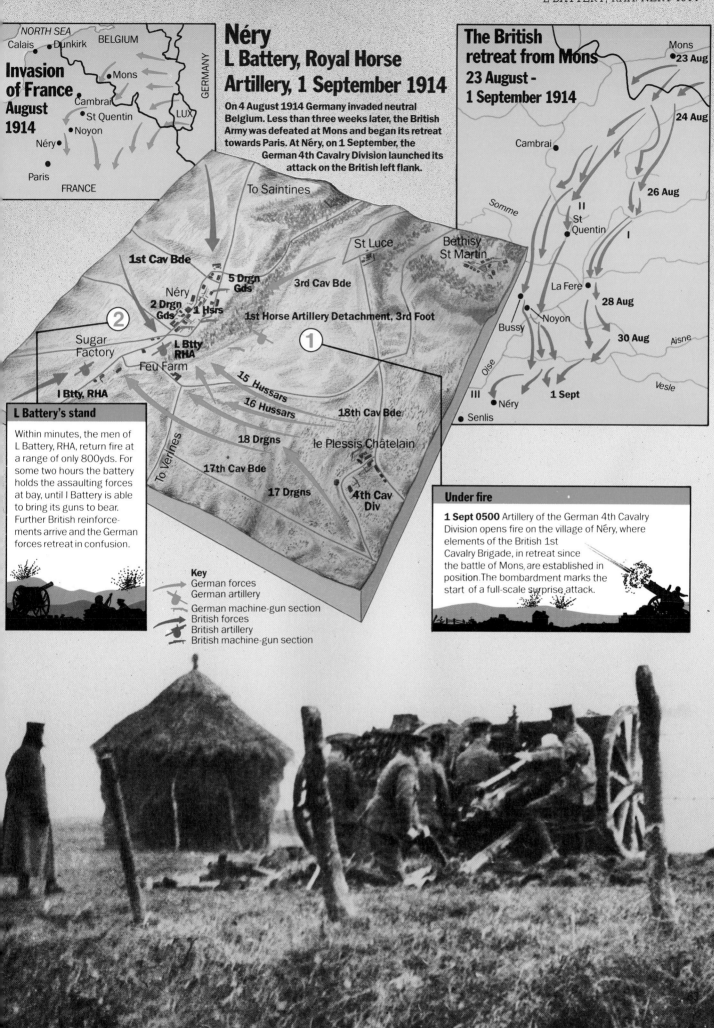

Invasion of France August 1914

NORTH SEA · Calais · Dunkirk · BELGIUM · Mons · Cambrai · St Quentin · Noyon · Néry · Paris · FRANCE · GERMANY · LUX

Néry
L Battery, Royal Horse Artillery, 1 September 1914

On 4 August 1914 Germany invaded neutral Belgium. Less than three weeks later, the British Army was defeated at Mons and began its retreat towards Paris. At Néry, on 1 September, the German 4th Cavalry Division launched its attack on the British left flank.

The British retreat from Mons
23 August – 1 September 1914

Mons · 23 Aug · 24 Aug · Cambrai · Somme · St Quentin · 26 Aug · II · I · La Fere · 28 Aug · Bussy · Noyon · 30 Aug · Aisne · Oise · III · Néry · 1 Sept · Vesle · Senlis

To Saintines · St Luce · Bethisy St Martin · 1st Cav Bde · 5 Drgn Gds · 3rd Cav Bde · Néry · 2 Drgn Gds · 1 Hsrs · 1st Horse Artillery Detachment, 3rd Foot · ② · Sugar Factory · L Btty RHA · Feu Farm · I Btty, RHA · ① · 15 Hussars · 16 Hussars · 18th Cav Bde · 18 Drgns · le Plessis Châtelain · To Verines · 17th Cav Bde · 17 Drgns · 4th Cav Div

L Battery's stand

Within minutes, the men of L Battery, RHA, return fire at a range of only 800yds. For some two hours the battery holds the assaulting forces at bay, until I Battery is able to bring its guns to bear. Further British reinforcements arrive and the German forces retreat in confusion.

Key
German forces
German artillery
German machine-gun section
British forces
British artillery
British machine-gun section

Under fire

1 Sept 0500 Artillery of the German 4th Cavalry Division opens fire on the village of Néry, where elements of the British 1st Cavalry Brigade, in retreat since the battle of Mons, are established in position. The bombardment marks the start of a full-scale surprise attack.

ears were bleeding because of the concussion, and I could not fire any more, so I left the seat and got a change by fetching ammunition.'

Darbyshire moved out just in time, as his replacement, Lieutenant Campbell, was not to last long. A shell exploded under the 13-pounder's shield, blasting Campbell out of the gun seat and hurling him through the air. Mortally wounded, Campbell lived for only a few minutes. Casualties mounted, and it was not long before Lieutenant Munday was hit in the right shoulder. However, he carried on using his left hand to bring up ammunition, until badly wounded in the right leg. Munday's wounds also proved fatal and he died a few days later.

Around 'F' gun, Captain Bradbury kept up his function as gun-layer alongside Sergeant Nelson who acted as range-setter, while Corporal Payne,

Gunner Darbyshire and Driver Osborne were responsible for bringing up ammunition. Further explosions killed Payne and wounded Nelson. 'F' gun's rate of fire began to slacken, but the arrival of Battery Sergeant-Major Dorrell injected vital new life into the crew.

Bradbury then instructed Dorrell to take his place as gun-layer while he went back to organise the re-supply of ammunition, now running perilously low. After walking back a few yards Bradbury was felled, a shell having smashed both his legs about six inches below the trunk. Despite being in great pain, he continued to direct the gunfire, before a further wound incapacitated him. Gunner Darbyshire related Bradbury's last minutes:

'Though the captain knew that death was very near, he thought of his men to the last, and when

Below: On parade. The drivers of a 13-pounder prepare to move off for a pre-war display. In action, RHA batteries were expected to keep up with the cavalry, providing that arm with mobile fire support. Once the war became bogged down, the gun's flexibility could not compensate for its lack of punch against fortified positions. Bottom: One of the Néry guns. This 13-pounder, 'C', was silenced, after firing eight rounds, by a German shell that exploded directly over its muzzle.

THE 13-POUNDER FIELD GUN

The Second Boer War (1899-1901) had revealed serious limitations in Britain's artillery and, as a direct consequence, a new range of field guns was introduced, including the 13-pounder QF (quick-firing) gun for use by the Royal Horse Artillery. The 13-pounder came into service in 1904 and, 10 years later, it had the distinction of firing the first British shot of World War I.

As a cavalry weapon, the 13-pounder saw only limited service in France and Belgium

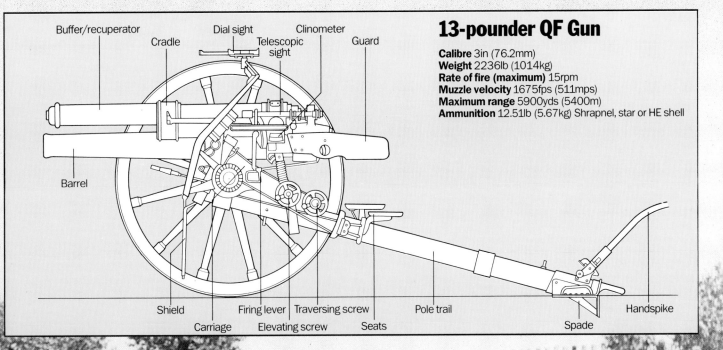

Buffer/recuperator · Cradle · Dial sight · Telescopic sight · Clinometer · Guard

Barrel

Shield · Carriage · Firing lever · Elevating screw · Traversing screw · Seats · Pole trail · Spade · Handspike

13-pounder QF Gun

Calibre 3in (76.2mm)
Weight 2236lb (1014kg)
Rate of fire (maximum) 15rpm
Muzzle velocity 1675fps (511mps)
Maximum range 5900yds (5400m)
Ammunition 12.51lb (5.67kg) Shrapnel, star or HE shell

once trench warfare set in at the end of 1914, especially so when the need for large calibre weapons became paramount. Nonetheless, the 13-pounder was a good light-field piece and was used extensively in the campaigns against the Turks in the Middle East, while on the Western Front a number of models were successfully adapted for use as anti-aircraft guns.

The 13-pounder had a calibre of 3in and fired a 12.5lb shell to a maximum range of 5900yds. Although the 13-pounder had a high-explosive capacity, its main ammunition type was shrapnel, an anti-personnel weapon consisting of a container holding 263 bullets, which, when detonated in the air by a small charge within the shell, flung out a deadly hail of lead or steel balls onto the target directly below. Against troops in the open, shrapnel was highly effective but a high-explosive shell was needed to deal with men behind earthworks and other similar defences. Consequently, large guns, able to carry out this role, were increasingly brought into service.

the pain became too much to bear he begged to be carried away, so that they should not be upset by seeing him, or hearing the cries which he could not restrain.'

Now only Dorrell and Nelson remained with the gun, firing the last few rounds amidst the hail of German shrapnel and high explosive. When the ammunition finally ran out, 'F' gun was silenced; Dorrell and Nelson slumped totally exhausted over the gun's breech. The war diary of the 5th Dragoon Guards relates the scene of devastation: horses had been blown completely apart – demonstrating the violence of a short-range bombardment – and amidst the carnage the three 13-pounders 'stood glaring grimly at the hostile guns'. Around 'F' gun, the dead and the dying lay among piles of spent cartridge cases.

The sacrifice of L Battery was not in vain, however, for the noise of the battle had alerted other British units, including I Battery RHA, which raced to the rescue of the hard-pressed British at Néry. The tables were turned, as the Germans had fully committed themselves to the assault on the village and were caught unawares by the sudden arrival of British reinforcements. One survivor of L Battery records the timely arrival of I Battery:

'Round they turned, stopping for nothing till they

Swamped by the massive numerical superiority of the German Army, the British Expeditionary Force was forced to make a withdrawal from Mons, a Belgium frontier town, in August 1914, and take up new positions along the Marne river. Bottom: Lance-carrying cavalry make their way south towards Paris. Although a small-scale action, Néry allowed the British Army to disengage from its pursuers and prepare a new defence line from where a counter-attack could be launched.

reached a ridge about 2000yds away. They unlimbered and got into action, and never was there grander music heard than that which greeted those of us who were left in L Battery when the saving shells of I Battery screamed over us.'

Totally surprised in their turn, the Germans retreated with the utmost haste; eight of their guns were left where they stood and the other four were recovered a short distance away.

The result of the battle at Néry, although a small action, was far reaching. The German 4th Cavalry Division took several days to recover and, as this formation was responsible for covering the outer flank of the German advance, its defeat allowed the British to retire and regroup unmolested, ready for the great counter-offensive on the Marne. Captain Bradbury, BSM Dorrell and Sergeant Nelson were awarded the Victoria Cross (Lieutenant Munday was recommended), Lieutenant Gifford received the Legion of Honour and Gunner Darbyshire and Driver Osborne were both awarded the Medaille Militaire. In 1926, as recognition of its epic fight, L Battery was awarded the battle honour 'Néry'.

THE AUTHOR Adrian Gilbert has edited and contributed to a number of military and naval publications and is author of a forthcoming book on World War I.

HEROES OF NÉRY

The outstanding bravery displayed by the men of L Battery during their defence of Néry was recognised by the award of three Victoria Crosses to the men who crewed 'F' gun: Captain Edward Bradbury (above left), Sergeant David Nelson (left) and Battery Sergeant-Major George Dorrell (right).

SLOPES OF DEATH

On the exposed mountain sides of the
Great Kabylia in Algeria, the
chasseurs alpins of the French Army
set out to destroy the guerrilla
fighters of the FLN – and found
themselves fighting a tenacious foe
that knew every inch of the
treacherous, rocky terrain

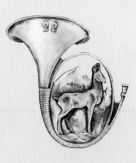

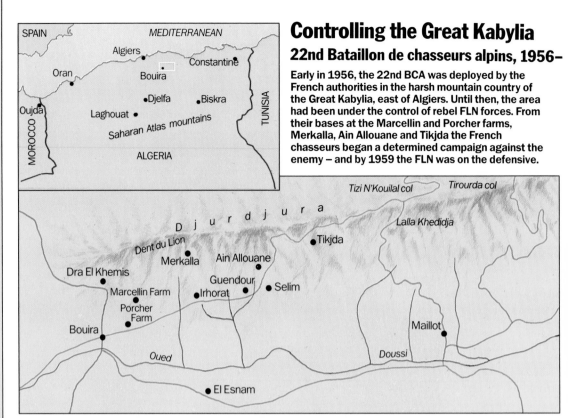

Controlling the Great Kabylia
22nd Bataillon de chasseurs alpins, 1956–

Early in 1956, the 22nd BCA was deployed by the French authorities in the harsh mountain country of the Great Kabylia, east of Algiers. Until then, the area had been under the control of rebel FLN forces. From their bases at the Marcellin and Porcher farms, Merkalla, Ain Allouane and Tikjda the French chasseurs began a determined campaign against the enemy – and by 1959 the FLN was on the defensive.

22ND BCA

The 22ᵉ Bataillon de chasseurs alpins (22nd BCA) fought in Algeria together with eight other battalions of French mountain troops. The 12th, 14th and 25th BCAs fought on the Tunisian frontier, while two formations, each of three battalions, fought in the Great Kabylia: the 6th, 7th, and 27th BCAs in one demi-brigade, and the 15th, 22nd and 28th in another. The 22nd BCA had its origin in the 22nd Battalion of foot chasseurs that was created as an 'alpine' unit in 1888 and based in the French Alps, at Albertville. The unit earned a distinguished record of service during World War I and then took part in the policing of the Rhineland.
In World War II, the 22nd BCA first defended the line of the Alps, but was then stationed in Alsace. Its men were part of the armies that tried in vain to halt the German onslaught into central France in 1940. Maintained as part of the army of Vichy France, the 22nd BCA was dissolved in 1942, but its members then formed the nucleus of two Resistance bands.
The unit was reformed in 1951, with its headquarters at Nice, and was sent to North Africa, first to Morocco and then to Algeria.
In May 1964, the unit was renamed the 22nd Group of Mountain Commandos (22nd GCA) but reverted to its original title of 22nd BCA in 1969. In 1976, however, the 22nd BCA was finally disbanded. Above: The horn, emblem of the chasseurs alpins.

UNDER THE burning sun of torrid summer, or in the powdery snow of icy winter, the battalions of French chasseurs alpins (mountain troops), made up almost entirely of young conscripts, held one of the harshest areas of the country throughout the Algerian War. In mountain scenery, sometimes reaching to over 2000m above sea level, they faced rebels who knew the country like the back of their hands, and had a well-earned martial reputation. But in a vast territory to the east of Algiers (corresponding to the 3rd Wilaya, one of the rebel administrative areas), these Frenchmen, from the French Alps of Savoy, Dauphiné and Provence, struck savage blows to their adversaries in the military wing of the FLN (Front de Libération Nationale or National Liberation Front).

Its complement increased in September 1955 by the arrival of reservists, the 22ᵉ Bataillon de chasseurs alpins (22nd BCA) left its traditional base at Nice for Morocco, where it was stationed near Oujda. A few months later the unit crossed the Algerian border to reach Michelet, on the northern slopes of the Djurdjura mountain chain. There, in the Great Kabylia mountains, they came up against a tenacious adversary, and had to deal with an escalating series of ambushes and hit-and-run raids. Battalion Commander Vuillemey, who had just taken over as the head of the 22nd BCA, soon received the order to transfer to Bouira, on the other side of the mountain chain, where he was to link up with the 6th BCA from Grenoble and Le Vercors. This area presented a particularly difficult task for counter-insurgency operations: a craggy mountainside rises to 2123m in the Dent du Lion or 'Lion's Tooth', while at the foot of the mountain wall a narrow plain slopes down to the ravine of the Wadi Ed Douss, through a series of hills covered by dense scrub. The Bouira military region as a whole spread over 300 sq km, and was populated by about 15,000 tribesmen living in 30 or so villages and hamlets.

This forbidding territory was divided up between the various elements of the 22nd BCA: PC and CCAS

(HQ and Command and Support and Services Company), were based at the Porcher farm at the edge of Bouira; the 4th Company at the Marcellin farm at the foot of the mountain; the 1st Company at Merkalla, below the Dent du Lion; the 3rd Company at Ain Allouane, over to the east at the same altitude; and the 2nd Company at Tikjda, a well-known mountain resort.

When the 22nd BCA arrived, this wild mountainous area was under rebel control. The nationalist guerrillas, described in official French communiques as HLL (hors-la-loi, 'outlaws'), had close control of the population through the OPA or 'politico-administrative organisation' that controlled several companies or katibas of more or less full-time troops. The chasseurs, therefore, had to operate on two levels: fighting the armed bands in the mountains while attempting at the same time to identify and destroy the political structure of the revolt in the villages. While companies and sections of chasseurs, stationed in small outposts, attempted to impose a military grid over this part of Great Kabylia, the rebels struggled to maintain their hold on the population through systematic liquidation of any fellow countrymen who appeared tempted to help the French. Soon after the arrival of the 22nd BCA, hostilities began to escalate – and the fight was on in earnest. The chasseurs were to learn that nothing could be taken for granted in this vicious guerrilla war.

The simple matter of opening up a road, normally an everyday operation, could easily turn into tragedy, for example. A little after 0800 hours on the morning of 16 May 1956, Sergeant Soulignac led a patrol of the 4th Company from the outpost of Tizi N'Djemaa, to clear the road up to the Tirourda col. To the west of the track, as Soulignac and his half dozen chasseurs pushed on, towered the huge mass of Azerout Tidjer, its stone eaten away by wind, ice and sun; to the east the view dropped away towards the villages of Summeur, Taklicht N'Ait Alsou and

Tirourda nestling in the valley below. It was a fine day: a light haze was resting on the tattered peaks. The chasseurs followed Soulignac at a good distance, taking either side of the road and ready to give each other covering fire in case of attack. Soulignac's six men were Corporal Aurensan and five chasseurs, nearly all of whom were from the Cote d'Azur: Dolcini, Lau, Aubert, Cavanna and Giordano.

The road clung to the side of the mountain and sometimes even burrowed into the rock through dark tunnels. The countryside seemed to breathe hostility, and all the patrol knew it harboured tenacious and resolute 'outlaws'. Soulignac moved into a bad stretch: the road was passing between two rock walls, like a slice into the stone massif. An archetypal site for an ambush – and all at once gunfire broke out.

The echo multiplied the shots – the whole mountain seemed to harbour rebels as bursts of fire fell into a hellish rhythm. Gunfire on all sides! The rebels were everywhere. Soulignac lay dead, his guts blown open by buckshot. Dolcini was wounded but returned fire...was hit a second time...and collapsed. The chasseurs at the tail of the patrol also came under fire. Corporal Aurensan and chasseurs Aubert and Cavanna were hit. But Giordano at the very back was unwounded and managed to break away to warn the command post. On the way, a sniper shot him in the shoulder, but he got through.

Suddenly, there were prolonged machine-gun bursts – from French weapons. Help had arrived and the rebels disappeared

Now only Lau was still okay. Returning fire with his Garand rifle from among the dead and wounded bodies, he managed to pick off some rebels who had prematurely moved down for looting. Cavanna and Aurensan also managed to fire despite their wounds. But they were now firing rather at random, because the guerrillas had got back to cover. Rebel fire became more accurate, homing in on whichever of the three chasseurs had just fired a shot. Lau was hit by something like a blow on the jaw – a ricochet or a splinter of stone. Soon his face was dripping with blood.

Lau screamed to give himself courage – 'You bastards!' – but now he realised that things were getting critical. He could see shadows shifting some 50m in front of him, rushing to new shelters in order to encircle the survivors of the ambush. Lau was desperate – but the range was short enough to use some of his rifle grenades. He fixed the special attachment to his Garand and fired some grenades towards the enemy. Explosions shook the mountain and deadly splinters of steel flew off the rocks.

The rebels kept their heads down, not moving, but Lau knew they were still there, waiting. His ammunition was low, and although Aubert crawled towards him with a sub-machine gun and some spare magazines, Lau's capacity to hold out until reinforcements came was limited – for the guerrillas were now opening up again. Then, suddenly, there were prolonged machine-gun bursts – from French weapons. Help had arrived from Tizi N'Djemaa, and the rebels disappeared with their wounded.

More or less alone, Lau, carrying on the old chasseur tradition, had held out in the most difficult circumstances. He was promoted to corporal, and decorated with the *médaille militaire* by the French president, René Coty.

The chasseurs continued their operations, learn-

ing all the while how to react to FLN attacks. On 12 March of the next year, 1957, a patrol from the 1st Company was fired on between Tanagount and Ait Haouari. Very soon the whole company was in contact with a large group of the enemy, and Vuillemey brought in all his other companies to encircle the area. The chasseurs were again successful: in a short time they had killed a further two of the enemy. This victory resulted in an about-turn in the attitude of the local population: some Berbers now announced that they were ready to set up 'self-defence sections', and were given arms. Even a local FLN political organiser joined one of these sections! Three hundred and fourteen shotguns were distributed, and this rallying of the Berber population was also marked by the opening of schools in the area. In addition, four groups of harkis, or native troops, were raised. Armed with automatic weapons, they were on constant alert.

Meanwhile, the war continued. On 23 December 1957 a rebel convoy of men and weapons, passing into the Bouira area from Tunisia, was intercepted by

Page 65 : Members of the 22nd BCA limber up for war in the mountains.

Top: General Jacques Faure with two of his chasseurs. In charge of the 27th Infantry Division, of which the 22nd BCA formed a part, Faure was a popular leader who had been an international athlete before World War II. Faure did not hesitate to make his political views plain, and was imprisoned for involvement in plots against the civil government during the agony of France's withdrawal from Algeria. Above: Clearing a road of mines – a nerve-racking and dangerous business.

the 4th Company in the wooded region of Talamine, south of the Wadi Ed Douss. The guerrillas lost a dozen men, but their presence remained strong. A three-company battalion of shock troops continued with attacks and ambushes. At the beginning of 1958, they ambushed a French artillery convoy east of Bouira, and the crews of two jeeps were slaughtered.

Faced with this renewed guerrilla activity, the 22nd BCA, under the new commander, Lieutenant-Colonel Giraud, had to hit back. Giraud called the 2nd Company in from its mountain post at Tikjda, and turned it into a hunter-killer unit based at Dra El Khemis, near Bouira. This counter-offensive was given renewed impetus when Charles de Gaulle came to power in France in 1958. He appointed General Maurice Challe as commander-in-chief, and a major pacification programme was soon under way. The Great Kabylia, one of the key centres of the revolution, was a vital target of Challe's strategy. The chasseurs combined military operations in the mountains with the opening of more schools; and various different units linked up in operations over the whole sector. One of the most crucial of these took place in October 1959 on the slopes of Lalla Khedidja.

Lalla Khedidja, named after one of the female saints of Islam, rises to 2309m on the southern flank of the Djurdjura chain. The area had always been a rebel refuge. The 2nd Company of the 22nd Battalion was part of 'Group C', one of four composite units created for this operation by sector command. The chasseurs' objective, at the centre of a large-scale manoeuvre, was the south face of Lalla Khedidja, together with the villages of Belbarra and Tala Rana. At 0600 on 2 October 1959, Group C left the operational HQ. The 2nd Company, under Captain Cha-

quin, leaving some troops behind with the packs, led the advance, arriving in Tala Rana just before 0900. There they found fresh tracks leading off towards Point 1566. The captain immediately launched his men onto the mountain: 'Second section goes to Point 1566, 1st to Point 1829; 3rd in the gully between; 4th in reserve with me.'

Sergeant Mausset's 2nd Section came under heavy fire from rebels entrenched among the rocks of Point 1566

It was an overcast day, with very low cloud. The summit of Lalla Khedidja occasionally disappeared from view. Suddenly the scouts leading the 1st Section came under fire from men in greenish fatigues at the head of the gully about 200m ahead. Sergeant Lebbe and three chasseurs were hit. Nearby, Sergeant Mausset's 2nd Section came under heavy fire from rebels entrenched among the rocks of Point 1566. The engagement rapidly became a hand-to-hand combat. One harki was killed, and then another, and a chasseur was hit. Young Mausset – a conscript – counter-attacked. His chasseurs drove off the immediate enemy, and then began to tend the wounded and retrieve dropped weapons.

The 3rd Section had moved off to the right, and it too came under fire from Point 1566. An Arab sergeant, Arfouni, and one of his harkis were hit; soon the whole company was engaged. Chaquin sent a group from his reserve section with a heavy machine-gun off to the right, hoping to outflank the enemy; another group was sent to reinforce the 3rd. The remainder stayed in place to hold a landing-site for the helicopter called to evacuate the wounded.

Above: The lonely posts that the chasseurs manned in the Algerian mountains were often vulnerable to surprise attack, but were essential for maintaining a constant French presence in the countryside.
Top right: Based in Tirilt outpost, these men successfully rallied Algerian tribesmen against the FLN – (from left), Captain Gibot, Commander Giraud and Captain Maviev. Centre right: Tribesmen who wished to set up 'self-defence sections' were issued with guns, if judged reliable by the authorities. Bottom right: the outpost 'Tour Sud' (South Tower) was one of the most isolated and inaccessible in Algeria. Of necessity, only the best men in the battalion were selected for its garrison.

Chaquin needed artillery support and air cover, although the cloud ceiling was very low; the rebels were everywhere. No sooner had the 60mm mortar of the command group been set up than it came under automatic fire and had to be moved; nearby, the captain's radio operator was hit in the thigh, and the aerial of his radio cut at the base by a bullet.

The guerrillas were firmly in control of the top of Point 1566, but when the re-sited mortar began working at them, hostile fire slackened. The respite was only temporary, however: the rebels regrouped and rained fire on the chasseurs from Point 1784.

Just after 1500 hours, French reinforcements arrived to help out the chasseurs. Chaquin directed them against Point 1784 with Mausset's section. The rebels, about 60 strong, charged down on the advancing French, with war-cries and screams of 'Don't shoot, comrades!' The French reinforcements, stunned by the force of the onslaught, took heavy casualties and began to fall back. Mausset's section was isolated, and after 10 minutes, the young sergeant had to withdraw his men to the cover of some rocks. Not until 1700 hours could Sergeant-Major Patrone's 1st Section come up to help; but then the lost ground was retaken, two rebels killed, and the French dead and wounded recovered. The enemy broke contact and the 1st Section took over the summit of the rocky ridge that was Point 1784.

It was a long, long wait till dawn, tending the wounded, with ears pricked for any sound of enemy activity

Chaquin now brought up his 3rd and 4th Sections; Mausset's battered 2nd Section was sent back to join the group with the supplies at Tala Rana. Chaquin decided to stick with the three remaining sections on 1784, as night approached. Soon it became very dark under the low cloud, and from time to time mist enveloped the chasseurs. Even in the icy cold they didn't dare light fires, and the packs were back at Tala Rana. It was a long, long wait till dawn, tending the wounded, with ears pricked for any sound of enemy activity.

Reinforcements arrived at first light, and the whole southern flank of Lalla Khedidja was occupied. The search on Point 1784 turned up four rebel dead and two rifles, and then the 2nd Company moved on down to Tala Rana and Saharidj, combing the slopes of Point 1566 as they went. The battle was the last major action fought in the Bouira sector by what remained of the notorious FLN Shock Battalion of Wilaya 3. At the start of 1958 it had counted nearly 500 men; but now it was down to 60.

Maraval de Bonnery succeeded Giraud as battalion commander on 5 November 1959; his task was as before, but by now there were only some 30 rebels left in the region, with just one light machine gun. This last automatic weapon was captured soon afterwards by parachutists who killed 14 rebels in one operation. The period of major military operations was finished: the chasseurs could concentrate their energies on 'hearts and minds' in the region, setting up medical centres, and new villages for the displaced population. They even organised a mountaineering training centre at the so-called Pigeon's Grotto near Ain Allouane.

To round up the last rebels, the 2nd Company took on a purely offensive role, and was involved in one of the last operations of the war, on the Djurdjura ridge. Under Captain Gaston, the chasseurs set out on

Tikjda, in the Great Kabylia, was a well-known Algerian mountain resort, but in the early 1960s, during the final years of the French presence in Algeria, it became the centre of French Army mountain warfare training, and the 22nd BCA became the headquarters company that ran the courses.

From 20 May 1962, the 27th Alpine Infantry Division undertook a series of intensive courses at Tikjda.

An officer of the 27th remembered the startling geography: 'Tikjda is like the altar of a temple or the crypt of a cathedral, its vast proportions carved from the rock; there is a nave stretching some 25km, and between the two passes of Tiz-Bou-El Ema and Tizi-N'Kouilal there rises a great vault, reaching 2300m high. The pillars of this great cathedral all have names – the Lion's Tooth, the Reygnier, the Jew's Hand.'

In these fantastic surroundings, the 22nd BCA had fought its long war against the FLN, and now trained hundreds of alpine troops in the rigours of mountain warfare.

patrol during the night of 17/18 July 1960. After two fruitless days, Gaston had decided to return to base, but in the late afternoon of the 20th, his look-outs noticed about 20 armed men moving in single file up a track. All the commando ambush-groups were alerted by radio. The rebels, totally unaware, came up to the level of Gaston's group, and the captain gave the signal for attack by firing an 81mm mortar shell at the tail of the enemy column. Despite the speed of their reaction, the rebels were trapped.

The chasseurs had killed none other than the renowned Si Salah, commander of Wilaya 4

Night was falling, and Gaston waited for first light before moving in. Meanwhile, his chasseurs maintained a tight cordon, and looked after a wounded comrade. At daybreak the French scoured the area and found six bodies. A burst of fire brought down another rebel as he tried to escape, and this was the first of the bodies to be identified. The chasseurs had killed none other than the renowned Si Salah, commander of Wilaya 4, the very leader who had once offered to surrender all his troops to de Gaulle. His escort was officers and NCOs, accompanying him to Tunisia. Si Salah was buried with full military honours at Bouira.

While the chasseurs had enjoyed military success, however, political events led to the final victory of the Algerian nationalists. President de Gaulle

decided, in spite of open opposition from certain elements in the army, to give Algeria its independence. The last military operations of the 22nd BCA took place, therefore, in a strange atmosphere, but the chasseurs continued to do their duty. By the start of 1962 the former rebel army of the Bouira region, which had once counted nearly 200 men, had been reduced to six men with only one sub-machine gun and six old bolt-action rifles between them. On 22 February this last resistance was wiped out in the cliffs to the south of the Wadi Ed Douss. Between Bouira and Maillot there remained only 30 or so poorly armed auxiliaries representing the FLN.

With the cease-fire, the 22nd BCA had the disheartening task of disarming harkis and the militiamen of the self-defence groups, closing down the schools they had opened, and abandoning some of their posts. At this point, the French high command found a channel to dissipate some of the bitterness among the troops: mountain warfare training became one of the key activities of the 22nd BCA. The first course opened on 20 May 1962 at Tikjda, now the mountain warfare training centre for the 27th Alpine Infantry Division. The 22nd continued to run the centre even after leaving Bouira to take up the occupation of the Rocher Noir and Reghaia com-

Top left and above: Following the declaration of Algeria's independence from France, the 22nd BCA turned its talents to organising mountain warfare training within their old battleground. Right: In the trackless Algerian highlands, pack mules were the only possible way of transporting goods to 22nd BCA's eyries in the mountains.

plexes. The battalion finally embarked for Marseilles on 31 January 1964. It was demobilised at Nice on 15 February, having lost 52 men in Algeria.

In May 1964, however, the unit reformed and was named the 22nd Group of Mountain Commandos (22nd GCA), but reverted to its original title in 1969. The 22nd BCA was finally disbanded in 1976.

THE AUTHOR Jean Mabire served with distinction in the *Commandos de Chasse* during the war in Algeria, and has written extensively on modern warfare. He is also author of several detailed studies of Nazi Germany's Waffen-SS.

BUSH FIGHTERS

South African Recce Commandos endure tough training to operate in hostile bush country

The first European settlers to arrive in South West Africa (SWA), renamed Namibia in 1968, were mostly Dutch, British and German. When tribal wars erupted in the 1880s, however, only Germany moved to protect her subjects and their property; by the 1890s the whole territory was under her control. The German regime was characterised by ruthless retaliations for hostile acts by the indigenous people. Thousands of Hottentots were slaughtered, and the Herero tribe of Bantus was reduced from 80,000 to 15,000 people after ferocious 'cleaning up' operations.

The Germans surrendered in 1915 to a military expedition from the Union of South Africa, and after World War I the Union was given full administrative powers in SWA under the Treaty of Versailles.

In 1946 it was proposed that the territory should become a fifth province in the Union of South Africa. Ignoring UN opposition, the Union put the proposal into effect. Finally, the UN voted to end South Africa's mandate to rule SWA, and in 1968 it renamed the territory Namibia as a preliminary to independence. When asked to withdraw from Namibia in 1971, South Africa refused.

Meanwhile, various nationalist movements had attempted to prise Namibia from South African hands. In 1965 an incursion of armed guerrillas into Namibia marked the beginning of a long conflict with the South African Special Forces that remains unresolved today. Above: The badge of the South African Special Forces.

THE RECONNAISSANCE Commandos, also known as 'The Recces', are the elite of South Africa's defence forces. Well trained in unconventional warfare, their main tasks are the destruction of strategic targets and reconnaissance, but they have many specialities and possess skills that are not found among ordinary members of the armed forces. The men are trained as individuals and in small groups to operate under difficult conditions and with little support. The Recces, like the military units created by other nations to undertake difficult and dangerous tasks – such as the American Green Berets, the British Special Air Service and the Rhodesian Selous Scouts – are a small unit and are highly selective, choosing only the very best men. Motivation must be good to begin with but, as with all elite forces, the men find real satisfaction in achieving the high standards set and then belonging to a group of men known to be of exceptional quality.

The Reconnaissance Commandos were formed by the South African government to deal with military threats at sea and on land which cannot be countered by conventional methods. To this end, their training is oriented towards the development of techniques of infiltration, reconnaissance and counter-subversion. The South African Special Forces are organised into naval and military branches and, with detachments at Durban, Langebaan, Phalaborwa and Pretoria, they are all controlled by Headquarters Special Forces at Voortrekkerhoogte.

The small naval element is made up of two components. The ship-borne element is trained to guard naval and mercantile shipping while in port (in order to do so personnel learn how to dispose of mines and explosives and are instructed in diving), and has a clandestine offensive capability against enemy craft. The second component is a marine group that can operate from surface warships and submarines. It can, for example, put combat teams ashore on the coastline in order to acquire information or strike at the enemy, later taking them off again in inflatable craft or letting them find their own way across country to link up with SA land forces. Naval Special Forces are not all parachute-trained but all are competent combat-swimmers.

They must perform 80 sit-ups in two minutes and eight pull-ups non-stop

The larger, military, group is composed of a highly-skilled fighting and training cadre of white officers and soldiers, with the equivalent of two battalions of black troops organised on a company basis. Some of these are ex-SWAPO guerrillas who, after years of military failure, have become disillusioned with the cause of African nationalism. Operating in their own environment and territory these black soldiers are exceptionally valuable for reconnaissance and the evaluation of enemy capabilities.

The white Reconnaissance Commandos, the kernel of the Special Forces, who also provide most of the officers in the black units, are either regular soldiers who have transferred from other arms or they are recruited directly on their entry into National Service. They must be South African citizens, volunteers, preferably unmarried, come from a good social situation, have a high leadership potential, have reached matriculation level in general education, have no criminal record, belong to a recognised church, and be exceptionally fit and strong. And all

Above: South African Special Forces personnel spill out from their Aérospatiale/ Westland SA.330 Puma transport helicopter. In the war against SWAPO guerrillas infiltrating into Namibia, the South Africans have made ample use of their air force to land troops in remote areas. Some of the men are carrying R-4 assault rifles, recognisable by their 'skeleton' stocks. Left: Armed with a 7.62mm FN MAG machine gun, a South African fighter prepares to set out.

Corps, in their case to eliminate impulsive and twitchy personalities.)

Having passed these tests, National Servicemen join the Recces' mainstream production line. The basic training lasts for three months; it includes an individual acclimatisation course of four weeks, a one-week selection course that tests the trainee under simulated battle conditions, and parachute training. After the preliminary course has been completed they then do eight months of special orientation for their role. They are instructed in the handling of mines and explosives, demolition methods, bushcraft, survival, boating, sailing, battle tactics, signals and first-aid. There is comprehensive weapon training on both their own and foreign weapons, and time is spent practising liaison with the air forces used in their support. Those who fail at any stage – and quite a high proportion of them do – are returned to their unit or, if they are National Servicemen who came straight to Special Forces training, they are sent to the technical service corps such as Signals, Ordnance, Intelligence or the Military Police.

...pounding around the parade-ground holding a 25kg cement block above the head...

The parachute course in particular is a very hard test. During the initial two weeks men do 10 periods every day, each of 40 minutes' duration, which include a high content of endurance training and body-building – as if the potential Recces haven't already proved the point! Apart from the toughening-up process they do longer and longer speed-marches, progressing by 5km stages up to 25km with full kit. Another element of this course is the dreaded pounding around the parade-ground holding a 'marble' – a 25kg cement block – above the head. The next phase is the actual parachute training, but only between 50 and 60 per cent of the young hopefuls who set out on the initial course of the Valkskermbataljon, the South African Parachute Regiment, graduate to that level.

Parachute specialisation in the post-war South African Army began in 1960 when 15 officers did basic training at RAF Abingdon. From this cadre grew the South African 1st Parachute Battalion based at Bloemfontein which now, understandably enough, provides a high proportion of the young regular Recces. Fundamental parachuting skills are the same worldwide but the Recces, having made their first static-line jumps from 150m, then go on to become experts in free-falling. Some then train to be high-altitude, low-opening (HALO) specialists.

The selection and basic training procedures of the white-cadre Recces have many similarities to those of the SAS, but later on their training and employment has greater affinity to that of the Selous Scouts, who also operated primarily in arid bush conditions. Indeed, many of the Selous Scouts who were recruited from the British or Rhodesian SAS when the unit was formed in 1974 went on to join the Recces after Rhodesia became Zimbabwe. The Recces consequently have a great depth of knowledge of urban as well as rural operations.

Once trained, the newly-qualified commando joins his operational unit; he will remain with it for the unexpired portion of his two years if he is a National Serviceman. Recces receive additional allowances

are stringently security-vetted before being accepted.

Before being drafted, all prospective National Servicemen are asked if they wish to volunteer for the Special Forces. If they do, they fill in a questionnaire which reveals the depth of their motivation and their qualifications. If the answers reveal that they are above average physically, psychologically and intellectually, they are entered for a special pre-selection examination and undergo rigorous tests which eliminate those who are unable to achieve the very high standards demanded. They must, for example, perform physical feats which are beyond the capabilities of most young men without special work-up training – 80 sit-ups in two minutes, 50 press-ups non-stop, eight pull-ups non-stop, running 200m carrying a man in a fireman's lift in one minute, 5km cross-country in 20 minutes, and a 15km route-march carrying a 30kg load in two hours. There are also psychometric tests to determine their character potential, in particular whether they will be able to stand the strains of prolonged operations behind enemy lines. (Similar psychological tests are used during the initial screening process of bomb-disposal operatives in the British Royal Army Ordnance

Above left: Working up to full fitness in preparation for his parachute jumps, a reconnaissance commando suffers the gruelling physical test of pounding a parade-ground holding a 25kg 'marble' above his head. Above right: A member of the South African Special Forces blacks up for an operation in one of the Namibian tribal homelands. Below: The Ratel 90 Fire Support Vehicle, an Eland 90 armoured car fitted with a 90mm gun. Right: A swarm of South African helicopters fly west into Namibia.

78

Cross-border operations
South African forces in Angola, 1978–1984

In 1978, as SWAPO infiltration into northern Namibia escalated, South African forces launched Operation Reindeer – the first of a series of cross-border raids against SWAPO bases inside Angola. In 1982 SWAPO's Eastern Front HQ was located near Mupa and raided by Reconnaissance Commandos, inflicting serious damage on the insurgents' logistics and command structure.

Key
SWAPO infiltration routes

1 Reindeer, May 1978

A battalion-size airborne force attacks a SWAPO camp near Cassinga, some 250km inside Angola. Further sweeps are carried out nearer the border.

2 Saffron, August 1979

After a SWAPO rocket attack on Katima Mulilo, South African forces mount a reprisal raid into southwest Zambia, virtually ending insurgent operations in the Caprivi strip.

3 Sceptic, June 1980

South African mechanised forces attack SWAPO bases in the 'Smokeshell' area, killing over 300 insurgents and capturing large quantities of equipment.

5 Protea, August 1981

A mechanised force crosses into Angola, advancing to Humbe. Another force secures SWAPO Northwestern Front HQ at Xangongo and takes out the HQ at Ongiva.

9 Askari, December 1983 – January 1984

Following reports of a SWAPO build-up, South African security forces mount a pre-emptive raid. The operation severely disrupts SWAPO logistics and organisation.

4 Klipkop, August 1980

After a series of insurgent operations in Kaokoland, a force is helicoptered into the village of Chitado to destroy the SWAPO base there.

8 Mebos, July – August 1982

After a protracted operation in SWAPO's Central and Eastern Front area, the Eastern Front HQ is located near Mupa and taken out in a lightning strike by Recce Commandos.

7 Super, March 1982

A small South African force is inserted by helicopter in the vicinity of a SWAPO camp near Cambeno. In an action lasting eight hours some 200 insurgents are killed.

6 Daisy, November 1981

A South African force crosses into Angola and attacks the SWAPO base at Bambi. The force pulls back after carrying out operations in the area for some three weeks.

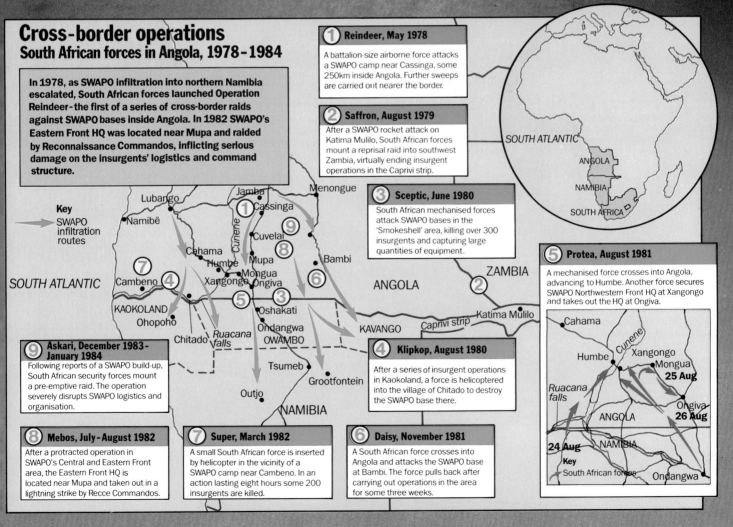

SOUTH ATLANTIC

ANGOLA

NAMIBIA

SOUTH AFRICA

Lubango • Jamba • Menongue
Namibé • ① Cassinga
Cahama ⑨
Humbe Cuvelai ⑧
⑦ Mupa
Cambeno ④ Mongua Bambi
SOUTH ATLANTIC Xangongo Ongiva ⑥ ANGOLA ② ZAMBIA
⑤ ③ Katima Mulilo
KAOKOLAND Oshakati Caprivi strip
Ohopoho Ondangwa KAVANGO
Chitado Ruacana OWAMBO
falls
Tsumeb
Outjo Grootfontein
NAMIBIA

Protea detail map:
Cahama
Humbe Cunene Xangongo
Mongua
25 Aug
Ruacana falls
Ongiva
ANGOLA **26 Aug**
24 Aug NAMIBIA
Key
South African forces
Ondangwa

for having passed their stringent training, and they also get extra pay during active operations. If a Recce reaches the rank of sergeant he may apply to be considered for promotion to officer status. At the end of the service period both regulars and National Servicemen are transferred to the 2nd Battalion of the Reconnaissance Commandos, the Citizen-Force unit in which they can continue to use their specialist skills until their reserve commitment has ended. Young National Servicemen can apply to become regular soldiers.

Since 1981 they have been involved in a multitude of clandestine operations

The Recces were employed repeatedly during the very protracted conflict with SWAPO guerrillas in Angola. They took part in the 1975-76 'Angola March' and, in 1980, in the big anti-terrorist operations 'Sceptic' and 'Smokeshellbasis'. In 1981 they were involved in 'Protea' and 'Daisy', and since then in a multitude of clandestine short-duration, hit-and-destroy attacks. Their targets were reached variously by parachute, on foot, in vehicles, by helicopter or across water. When deployed in the bush the white-cadre Recces use the base of a conventional front-line support unit during protracted operations, but in short-lived 'take-out' attacks they parachute in direct

Below: In the savage beauty of the southern African landscape, a patrol returns to base, having scoured the bush for signs of SWAPO guerrillas.

from their home base and then extract themselves when their mission has been accomplished. The black troops operate as self-contained units or in small groups for the duration of the operations and are then withdrawn to their base camps.

Operating in areas which are entirely populated by blacks the white cadre is careful to disguise its race, and for this reason white Recces on active service black-up all exposed flesh. In some instances pseudo-teams will go into action disguised as guerrilla groups and carrying Soviet weapons such as the AK47 or the RPD machine gun. Some Recces use the 7.62mm FN MAG machine gun, which sprays a lot of lethal hard stuff at close quarters, but most carry the FN/FAL NATO rifle with a folding stock. At the belt they carry a blackened stiletto, a water bottle, two days' concentrated rations, ammunition pouches and immediate first-aid.

The Recces take great pride in belonging to the fittest, toughest and most highly-trained outfit in the South African Defence Forces. As individuals they are highly-specialised and very formidable fighting men, and as units they enjoy an unequalled reputation for successful military operations in the formidable conditions of the African bush.

THE AUTHOR Brigadier Peter Macdonald served with the Royal Army Ordnance Corps, specialising in terrorist bomb disposal, in Cyprus, Aden and Northern Ireland. He has published several books on the subject including *Wide Horizons, Exit* and *One Way Street.*

LEATHERHEADS

THE RED BRIGADES

The far left-wing Red Brigades first came to the notice of the Italian public in August 1970. At first they restricted their activity to passing leaflets to workers at the SIT-Siemens factory in Milan. Then, in December 1971, the organisation carried out its first bank raid to raise funds, and its first kidnapping took place three months later.

In 1974 the Red Brigades began a campaign against figureheads of Italian institutions. On 18 April they kidnapped a judge in Genoa, and later two neo-fascists were killed in a raid on a right-wing party's headquarters. The campaign moved to the level of organised political assassination in 1976 with the shooting of the Genoa Public Prosecutor and two bodyguards, and in 1977 the president of the Turin Association of Lawyers was killed and the deputy editor of a Turin newspaper was mortally wounded. The lives of two more judges were claimed before the Red Brigades carried out their best-known act, the kidnap and murder of ex-Prime Minister Aldo Moro (below) in 1978.

At the time of the Moro killing, the Red Brigades were organised into three groups. The first was a nucleus of up to 500 full-time members, working on a salary of approximately £200 a month. The second consisted of about 1000 members who participated only occasionally in the organisation's operations. Third, several thousand more members were enlisted to help the Red Brigades by carrying out such minor tasks as renting apartments and carrying messages, thereby shielding core members from undue risk.

Held hostage by the Red Brigades, General James Dozier of the US Army was rescued by Italy's crack anti-terrorist squad, the Nucleo Operativo Centrale di Sicurezza

ON THE COLD morning of 28 January 1982, the power plant of a bulldozer sprang into life on a construction site near the Via Pindemonte in Padua, northern Italy. Moments later, shoppers in the suburban street were shocked to see heavily armed men, wearing ski masks or balaclavas and bullet-proof vests, rushing from a van and pounding up the stairs of No. 2 Via Pindemonte, a residential apartment block with a supermarket on the ground floor. It looked as though Italy was about to suffer yet another act of political terrorism.

In this instance, however, appearances were deceptive. Within a first floor apartment were members of Italy's most feared terrorist organisation, the Brigate Rosse (Red Brigades), who were standing guard over their latest hostage, US Brigadier-General James Dozier. Their senses dulled by six weeks of monotonous waiting, they failed to detect the rapid approach of the masked gunmen over the roar of the bulldozer. Little did they realise it, but the long string of successes enjoyed by the Red Brigades was about to be brought to an end by men of Italy's top anti-terrorist unit who were poised to come crashing through their door.

The organisation known as the Red Brigades emerged as one of Europe's pre-eminent terrorist groups in the early 1970s. Similar in philosophy and methods to Germany's Red Army Faction (the Baader-Meinhof Gang), the Red Brigades were an extreme left-wing group with the espoused aim of introducing communism through the use of terrorism. Later in the decade the Red Brigades became famous for their campaign of terrorism against gov-

ernment officials, corporate executives and journalists. Their speciality was 'knee-capping' (crippling their victims by shattering their knee-caps), and they also committed several assassinations. Acts of arson were also carried out against the Fiat motor company, the executives of which were also prime targets for shootings and kidnappings. In 1978, while leaders of the Red Brigades were standing trial in Turin, the former Italian Prime Minister Aldo Moro was kidnapped and murdered by the organisation. This act so incensed the Italian people that the authorities proceeded to take draconian action against the nation's terrorist groups, and a large number of arrests were made.

Possibly as a way of demonstrating their continued potency, the Red Brigades took General Dozier hostage on 17 December 1981. The second-highest ranking US Army official in southern Europe, Dozier was seized in the late afternoon at his riverside home in Verona by four men masquerading as plumbers. His wife was overpowered and bound, and Dozier was led away after a blow to the head with a pistol. As it was later learned, the party sped to Padua, 50 miles from Verona, where Dozier was installed in the Via Pindemonte apartment. In order to ensure that he would be unable to recognise the terrorists if released, or identify the apartment where he had been held, Dozier was forced to spend long hours in a tent erected in the room, with his left foot and right hand chained to the central pole. His captors' faces were covered on the occasions when he was let out. The room's blinds were kept drawn and the tent was permanently lit by a 40-watt bulb, making it impossible to distinguish day from night. In order to mask their conversation, the terrorists

forced Dozier to listen to loud rock music through headphones, an expedient which caused permanent damage to his hearing.

The kidnapping of General Dozier, the most spectacular act of the Red Brigades since the kidnap of Aldo Moro, took both the Italian authorities and NATO by surprise. Although the idea of selecting a military figure as a target had been canvassed in several of the Brigades' 'strategic resolutions', previously all of their victims had been Italian. Possibly the terrorists were unprepared for the consequences of their act, for the kidnapping of a high-ranking NATO officer spurred Italy's anti-terrorist forces into an all-out effort against the Red Brigades. Over 6000 Italian law-enforcement personnel were deployed in the search for him, and experts were flown in to assist from several countries that had their own sophisticated anti-terrorist squads. From the US came six members of the Counter-Terrorist Joint Task Force (CTJTF), special agents were sent from Bonn, and advisers arrived from Britain.

Meanwhile, the terrorists found their captive, a former officer of the US Rangers, a very hard man. On 18 December a message had reached the world: 'We claim the kidnapping of NATO hangman James Dozier last night ... He is held in a people's prison and

Page 80 : An awesome display of Italian anti-terrorist firepower. Left: A NOCS member abseils into action. Above: A NOCS team practises insertion by helicopter. Above left: Masked NOCS commandos are photographed as they set out on a security mission.

will face proletarian justice.' Although the 50-year old Dozier was kept in uncomfortable conditions for 42 days, he emerged a few pounds lighter but undaunted. His interrogators had learned very little from their 'assassin and hero of the American massacres in Vietnam', and Dozier presented them with no tapes or statements to help further their cause.

The manhunt for Dozier increased in intensity as December drew to a close. Dozens of terrorists were rounded up and large stockpiles of weapons and explosives were discovered; one was found in four suitcases buried four feet underground in the mountains 12 miles north of Treviso. Machine guns, anti-tank grenades, shotguns, hand grenades, thousands of rounds of ammunition, explosives and fuzes were seized. The loss of men and equipment gradually bit into the Red Brigades' operational effectiveness, and further arrests would follow the release of Dozier.

On 4 January 1982, plainclothes men arrested two men as they drove through central Rome, heavily armed and apparently en route to a kidnapping. Five days later, police charged into an apartment in Rome and caught Giovanni Senzani, a former university professor turned Red Brigades column leader, with an arsenal of weapons. Then five more alleged terrorists were arrested in the countryside north of Rome by police hunting the killers of two young policemen in a bank robbery in Siena. Information received from one of these sources led the police to Padua, where they quietly identified the flat in the Via Pindemonte.

Once the apartment had been singled out, the Italians acted cautiously. A night raid was ruled out for fear of a policeman mistakenly shooting Dozier in the dark, and a dawn attack was also discounted as the terrorists would notice the increase of people in the street. The time of the assault was set for late morning. The men chosen to carry it out were 10 hand-picked police commandos from the shadowy Nucleo Operativo Centrale di Sicurezza (NOCS) – the Central Operative Nucleus of Security.

NOCS commandos were chosen from the fittest and most intelligent volunteers in the police

Part of Il Corpo delle Guardie di Pubblica Sicurezza (the Corps of Public Security Guards), NOCS consisted of about 50 highly trained police officers. Known as the 'Teste di cuoio', or 'Leatherheads', after the leather helmets, designed to give full protection to the face and neck, that they wore in action, NOCS commandos were drawn from the fittest and most intelligent volunteers in the police. A selection course eliminated all but the top candidates, whose training was then conducted along lines similar to those of the British SAS, the French GIGN and the German GSG-9. Taking place at the Abbasanta Police Training Centre in the hills of Sardinia, the training included hand-to-hand combat, high-speed driving and combat shooting with the Beretta Model 12 sub-machine gun and other weapons. Also taught were breaching techniques and the use of special-purpose munitions such as stun grenades, rappelling, terrorist tactics and psychology, electronic surveillance and other skills required of a top-notch hostage rescue unit (HRU).

On 28 January 1982 the men of NOCS were given a classic opportunity to put their specialised training into effect: the assault on the apartment in Via Pindemonte. In the late morning hours of that day, the members of NOCS and other police officers began to

Far left, below: Hand-picked volunteers of the Nucleo Operativo Centrale di Sicurezza (who wear the Italian police badge inset at top) train in all aspects of hostage rescue operations. Far left: The apartment in Via Pindemonte (arrowed) where General James Dozier was held for 42 days. Left: This photograph of General Dozier, bearded after weeks of captivity and holding a message from the Red Brigades, with the symbol of the terrorist organisation behind him, was later shown to be a montage. Below: General Dozier after his release from captivity. Bottom: Two NOCS commandos hustle the general to safety after their spectacular four-minute raid on the Red Brigades' hideout.

BRIGADIER-GENERAL JAMES DOZIER

For a hostage snatched from the ranks of the US Army establishment in Europe, the Red Brigades could hardly have chosen a tougher individual than James Lee Dozier. A West Point graduate, Dozier had served in the US Army for over 25 years and was known as a soldier's soldier by his fellow officers. About 5ft 9in tall, he had gained a reputation for personal hardness in the Vietnam War, where he was awarded the Silver Star and three Bronze Stars for bravery, along with the Purple Heart for being wounded in combat. He served with an armoured cavalry regiment in Vietnam, and following the war he was attached to armoured units in West Germany. He also commanded a brigade of the 2nd Armored Division at Fort Hood, Texas, as well as holding a number of staff posts.

In June 1980 General Dozier was appointed Deputy Chief of Staff at the Verona headquarters of NATO's land forces in southern Europe, a position that made him the second-highest-ranking US Army official in the area. Described as wiry and in top physical shape, Dozier found the worst aspects of his captivity to be boredom and withdrawal from his daily exercise regime. Given reasonably well-balanced food during his 42-day ordeal, his choice for the first meal out of captivity was an all-American favourite – a cheeseburger, french fries and a Coke!

Among the many commendations he received for his bearing during the crisis was one from President Ronald Reagan:

'The same courage and resolve that James Dozier demonstrated on the battlefield in wartime have seen him through this new test with flying colours. His country and our allies can be very proud of this gallant man...his rescue is welcome news for all those who believe in the rule of law and the defence of our free institutions.'

Above left: Fully restored after his ordeal and back in the uniform of a US Army general, James Dozier recounts his experiences at a Pentagon press conference on 4 February. Behind him is an artwork of the tent in which he was held.

move into position near the apartment building. Civilians working in the area were quietly evacuated shortly before 1130 hours in preparation for the operation, and the engine of the bulldozer was started up to supply covering noise. Shortly after 1130 a van pulled up in front of the apartment and the 10-man NOCS assault element rushed out and into action. Young men in blue jeans, police agents in plain clothes, immediately blocked the exits of the supermarket below the apartment to prevent any shoppers from wandering into the line of fire, for it was feared that the terrorists might gain an opportunity to use their weapons. (One woman later related that she had called home to say that she was being held prisoner during what she thought was a robbery!) The scene was set.

At 1136 the commandos, wearing body armour and carrying Beretta sub-machine guns, stormed up the stairs to the first floor. Their leader, a powerful man who was a competitive weightlifter, took out the

door of the apartment with a single blow. Rushing in, the commandos found themselves face to face with a terrorist holding bags of groceries, who had just returned from the shop below. An NOCS commando dropped him with a karate chop before he could react.

In the first room to the right, General James Dozier sat bound and gagged in the blue tent, surrounded by four terrorists, two men and two women. Hearing the commotion outside, one of the men raised a silenced pistol at Dozier to carry out his execution. One of the assault team was quicker, however, and he used the butt of his M-12 to drop the terrorist with a blow to the neck. The man crumpled to the floor, and the other three surrendered without resistance. In less than 90 seconds the NOCS team had deprived the Red Brigades of their prize. Relieved of his gag, Dozier showed his admiration – 'Wonderful!' The general was quickly evacuated, while the NOCS members remained under cover to trap any terrorist who might unwittingly travel into the area.

Arrested in the assault were Emanuela Frascella,

Antonio Savasta, Emilia Libera, Cesare di Lenardo and Giovanni Ciucci. Savasta was an important Red Brigades leader who was suspected of having played a key role in the kidnapping of Aldo Moro. In the aftermath of the assault several more key members were also arrested in Verona, Padua and Mestre, near Venice. The Italian public, meanwhile, were quick to show their appreciation of the 'Leatherheads', who had won such a clearcut victory over terrorism in their country.

Unfortunately, the excellent reputation of the Nucleo Operativo Centrale di Sicurezza suffered from subsequent events. Five of General Dozier's rescuers were later convicted of torturing members of the Red Brigades and, during the last few years, the unit has been eclipsed by a rival formation. Drawn from the ranks of Carabinieri, the Groupe Interventional Speciale (GIS) – the Special Intervention Group – has superseded NOCS as Italy's primary anti-terrorist unit.

Based near the town of Lavarno, GIS consists of 46 volunteers from the Carabinieri under a major. Selected from parachute-qualified personnel (the Carabinieri has its own parachute unit), GIS receives intensive physical training. Every day GIS members complete a five kilometre cross-country run, a two kilometre swim, and hand-to-hand combat training.

There is also marked emphasis on shooting training, including the use of an SAS-style 'killing house', in which members of the unit act as hostages during live-fire exercises. Members are trained as snipers in order to work in conjunction with assault teams, while demolitions and breaching techniques, rappelling and other entry methods are rehearsed until the men work faultlessly as a team. Training in anti-hijacking techniques also takes place at Rome's Leonardo da Vinci Airport. GIS members carry the 9mm Heckler and Koch MP5 sub-machine gun rather than the Beretta Model 12 used by NOCS, and are trained in the use of gas and other special-purpose weapons.

Despite the fact that GIS has now usurped some of NOCS's glory, and despite the scandal which later resulted in some NOCS members being sent to prison, it was still the 'Leatherheads' of NOCS who carried out Italy's most dramatic rescue against the Red Brigades and who, therefore, struck the most significant blow in Italy's war against terrorism. As a result, NOCS's place in the history of elite anti-terrorism units is assured.

THE AUTHOR Leroy Thompson served in Vietnam as a commissioned officer in the USAF Combat Security Police. He has served with VIP protection units and advises on counter-terrorism and the training of hostage rescue units.

Above: Formed in 1978 after the murder of ex-Prime Minister Aldo Moro, the Nucleo Operativo Centrale di Sicurezza took orders only from its commander, Gaspare de Francisci, the man appointed by the Italian government to fight terrorism. Here, members are shown with an Agusta-Bell 212 helicopter.

In Delta Force, the United States Army has a unit capable of tackling the growing threat of international terrorism

SOON AFTER 0200 on 18 October 1977, the crack West German anti-terrorist unit, GSG9, stormed a hijacked Lufthansa jet airliner on the ground at Mogadishu in Somalia. The assault was a triumph for Ulrich Wegener's fledgling force: three terrorists were killed and the aircraft was taken intact without the loss of a single hostage. A month later, on 19 November, the United States Department of Defense finally activated its own elite counter-terrorist unit modelled, like GSG9, on Britain's 22nd Special Air Service (22 SAS) Regiment. Officially designated 1st Special Forces Operational Detachment – Delta, the unit quickly became known as Delta Force.

Backed up by other specialist units, Delta Force is the highly trained – and highly secret – spearhead of US counter-terrorist capability. Behind President Ronald Reagan's tough stance against terrorism lies the ability to call on Delta Force to carry out the kind of counter-terrorist assault pioneered by the British SAS and Israeli special forces.

Delta Force, with the pugnacious Beckwith as its commander, was activated on 19 November 1977

The need for specialised anti-terrorist units within the military structure was recognised first by Western European governments and by the Israeli armed forces, largely because terrorism is more conspicuous in Europe and the Middle East. Long before the Mogadishu operation, 22 SAS had added counter-terrorist operations to its brief. Similarly, West Germany's GSG9 was formed in the aftermath of a poorly co-ordinated and ill-fated rescue attempt by police marksmen during the 1972 Munich Olympics – resulting in the death of nine Israeli hostages held by a Black September terrorist group. By 1977, when GSG9 deployed to Mogadishu, the unit had the benefit of a long period of intensive training, in addition to SAS specialist weapons and advice.

Although European governments benefited from the lessons learnt during the 1970s and established units to counter the rising tide of terrorism, Pentagon bureaucracy and endemic inter-service competition for roles, resources and manpower delayed the formation of a specialist unit in the United States. These difficulties were exacerbated by the running down of the US Special Forces community during the mid-1970s in the aftermath of the Vietnam War. Although a number of American military planners began to recognise the need for a specialist anti-terrorist force along the lines of Britain's SAS, the political situation remained unfavourable. That the US got such a unit at all was, to a considerable extent, due to the efforts of one man – Colonel Charles Beckwith, who had been lobbying the Pentagon since 1963 for the formation of a special operations force.

Delta Force, with the pugnacious Beckwith as its

Designed to realise the potential of each recruit, the rigorous training programme at Fort Bragg (right) instructs operators in the techniques of rappelling (above right), river assault (top right) and the operation of selected vehicles (above far right).

DELTA

FORCE

COLONEL CHARLES
BECKWITH

'Chargin' Charlie' Beckwith
(above), served for three
years in the 82nd Airborne
Division before qualifying
as a para-trained officer. He
was posted to Fort Bragg,
North Carolina, to join the
Special Forces in 1958. Four
years later, as a Green Beret
captain, he was sent to
Britain for a year as part of
an exchange programme
between the British SAS and
the US Army's Special
Forces, joining A Squadron,
22 SAS, in June 1962. From
January 1963 Beckwith was
stationed in Malaya with this
squadron, gaining
invaluable experience in
British counter-insurgency
methods.

On his return to the United
States, Beckwith began to
campaign for an American
unit to be set up on SAS
lines, but his reports went
unheeded. He saw active
service in Vietnam, as part
of 5th Special Forces Group
in 1965 and 1966, and after a
further combat tour in 1968
he returned to the United
States. Following several
more postings in various
parts of Southeast Asia and
the Pacific area, Beckwith
was finally returned to Fort
Bragg in 1974 and appointed
commandant of the Special
Forces School.

From 1975 onwards,
Beckwith was repeatedly
consulted as the US Army
general staff began to take
an interest in the SAS
concept. He became heavily
involved in planning for a
new counter-terrorist unit
and finally, in August 1977,
Beckwith was relieved of his
Special Forces School
command to concentrate
entirely on setting up the
unit that came into being in
November as Delta Force.
Colonel Charles Beckwith
retired from the US Army in
1981.

After Operation Eagle Claw Delta's next major operation was in Grenada, as part of a task force sent to invade the island on 25 October 1983. In addition to elements from Delta Force, the Grenadian invasion force consisted of the 1st and 2nd Battalions of the 75th (Ranger) Infantry Regiment, together with elements of the 96th Civil Affairs Battalion and the 4th Psychological Operations Group (Psyops) – all of which form part of the US Army's 1st Special Operations Command (SOCOM) – plus units of US Navy SEAL commandos. After the SEAL commando teams had checked the beaches and harbours before seizing the radio transmitters at Radio Grenada and the Governor's residence, the Rangers spearheaded the main assault.

The two battalions made a low-level parachute drop on Point Salines airfield and secured the area after two hours of hard fighting. Delta Force arrived during the Ranger airdrop and was tasked to attack Richmond Hill prison – intelligence had indicated the presence of political prisoners within the compound.

However, the attack was a conventional daylight assault on a heavily defended enemy position – an operation that could not possibly be expected to exploit the specialist counter-terrorist skills of Delta Force. In the event, intelligence proved to be faulty and there were no prisoners at Richmond Hill. Nevertheless, the Cuban defenders were well dug in and Delta pulled back under heavy fire.

For the remainder of operation Urgent Fury, Delta Force carried out a number of quick-strike attacks behind enemy lines. Delta was supported in this capacity by helicopters from Task Force 160 of the 101st Army Air Assault Division.

first commander, was activated on 19 November 1977 and organised into SAS-type squadrons, composed of 16-man troops divided into highly flexible four-man patrols. The formation was to be raised by careful selection of volunteers from other units – predominantly but not exclusively from the Special Forces – who would then be carefully trained for the counter-terrorist role.

Delta operators must be able to hit their targets with 100 per cent accuracy at 600yds

When the Delta project was activated, the first priority was to put together a rigorous selection and training course. Some 30 volunteers were selected to form the nucleus of the new unit. Along with others possessing experience of SAS selection methods, Beckwith organised a gruelling physical training test that included a 40yd inverted crawl in 25 seconds, a two-mile run, a 100m fully-dressed swim, and a speed-march over 18 miles. About half the candidates in the first group were eliminated by this stage. An SAS-type selection course followed, consisting of a lone endurance run over mountainous and thickly-wooded terrain in full kit – including a 55lb rucksack. Each man had to push himself to the limit: the time required to make each rendezvous was never revealed, and the course continued for several days and nights. Finally, the survivors were put through a thorough psychological screening to assess their mental abilities and motivation. This initial cadre acted as instructors, staffing a second selection course held along similar lines in January 1978. By the end of April 1978, five selection courses had been run and, out of a total of 264 volunteers, 73 men had been chosen for further training – enough to form Delta Force's A Squadron.

The next stage, once a volunteer had been accepted, was an intense 19-week Operator's Course at Delta's headquarters – the secure stockade at Fort Bragg, North Carolina. Smallarms training formed a major part of the course: Delta operators must be able to hit their targets with 100 per cent accuracy at 600yds, and to achieve this extraor-

dinary level of marksmanship they are expected to spend three to four hours a day shooting. Weapons include modified 0.22mm Berettas designed for use in confined spaces such as aircraft, 0.45in Colt pistols, shotguns, specially built Remington 40XB sniper rifles, M3A1 and M16 rifles, M60 and Heckler and Koch HK21 light machine guns, Heckler and Koch MP5 sub-machine guns, and M79 and M203 grenade launchers. A range of specialist devices, including C4 plastic high-explosive charges and stun-grenades are also in use, and form a vital part of Delta's armoury. A 'shooting-house' similar to that used by the British SAS was built in the Delta compound, allowing realistic assaults to be practised. Room-clearing exercises involved filmed sequences of hostages and terrorists – the trainee would enter the room as part of a team and be responsible for distinguishing terrorist from hostage before opening fire. Aircraft assaults were carried out inside mock-up cabins. By early 1979, as a result of this specialised training, Delta Force was nearing operational readiness – Beckwith had originally forecast an initial working-up period of two years. Strength was up to two squadrons, an exchange programme with the SAS was underway, and contacts were established with GSG9, the French GIGN and the Israeli special forces.

In its first two years of existence, Delta had not been tried in action. Now came the force's first opportunity to carry out an operation. On 4 November 1979, some 400 militant Iranian students stormed the United States embassy in Tehran, the Iranian capital, and took 66 American citizens hostage. By 20

**The US rescue plan
Operation Eagle Claw**

Key
— Route of helicopters
— Route of C-130 transports
- - - Route of C-141 transports

November, 13 of the hostages had been released but the remaining 53 were still being held within the embassy compound. Despite his public disavowal of a military solution, President Jimmy Carter had considered the use of military force to solve the crisis from the outset. Delta Force was ordered to develop a viable plan to retake the embassy by force and evacuate the hostages.

The main obstacles to a successful rescue were logistic. Delta Force itself had no integral logistic capability, and would have to rely on other units of the armed forces. Air transport – probably helicopters – would be needed to extract the hostages once they were freed, and it was therefore decided to use these same helicopters to insert Delta Force. The helicopters chosen were Sikorsky RH-53D Sea Stallions operated by Marine Corps pilots from USS *Nimitz* in the Gulf of Oman. To solve the problem of refuelling these helicopters during their journey into and out of Iran, a complex plan was developed involving Air Force fixed-wing tankers, transports and gunships, and a force of Rangers to defend two separate landing areas.

The final plan for the operation – known as 'Eagle Claw' – called for three MC-130 transports to fly the 118-man assault force to Desert One, a rendezvous some 300 miles southeast of Tehran. The transports would be followed by three KC-130 tankers. Meanwhile, eight Sea Stallions would take off from *Nimitz* and fly to Desert One. While the landing zone was

Below left: Three Sikorsky RH-53D Sea Stallion helicopters await orders to lift off from the deck of USS *Nimitz* following last-minute flight checks by the groundcrew (inset). These helicopters were both the key to success and the ultimate cause of Delta's failure during Operation Eagle Claw. Bottom right: As the hostages are assembled for the world's press in Tehran, anti-US feeling reaches frenzy level with the burning of American flags throughout Iran (below).

CARTER IS SUPPORTING THIS NASTY CRIMINAL UNDER THE PRETEXT OF SICKNESS

secured by a Ranger road-watch team, the helicopters would refuel and fly Delta Force on to a 'hide site' near Tehran, before moving off to a second hide site to await the call to move in and evacuate the hostages from the embassy compound. After a reconnaissance by Beckwith, Delta would proceed to Tehran by truck under cover of darkness on the second day of the operation. The assault plan called for three sections of Delta to act independently. While Red and Blue Elements seized different parts of the compound where the hostages were being held, White Element would seal off the embassy and cover the withdrawal. Air Force gunships would then provide air cover while the Sea Stallions evacuated the hostages. It was agreed that a minimum of six Sea Stallions would be needed for this stage of the operation. The hostages would be flown to Manzariyeh airfield, some 35 miles south of Tehran, where a force of Rangers would be guarding three C-141 Starlifters ready to fly the hostages and their rescuers out of Iran. The last Rangers would leave by C-130 after destroying the helicopters on the ground.

The assault force finally withdrew – defeated by equipment failures outside their control

After months of intensive training and rehearsal, Delta was deployed to Egypt on 21 April 1980. Three days later, the go-ahead was given and Delta was airlifted to the island of Masirah in the Gulf of Oman. At 1630 on the 24th, the assault force boarded three troop-carrying C-130 transports, and 90 minutes later the first transport took off and set course for Desert One, arriving safely some four hours later. The Ranger road-watch team raced out of the C-130 to secure the landing zone. However, soon after the landing a bus, and later two fuel tankers, drove into the landing area. The bus passengers were taken prisoner, but one of the fuel tankers had to be taken out. One vehicle escaped, representing a dangerous security risk. The remaining C-130 transports and tankers arrived and the assault force waited for the arrival of the Sea Stallions.

The helicopters had left *Nimitz* on schedule at 1930, but two Sea Stallions failed to reach Desert One. Since a minimum of six were required to complete the operation, there was now no margin for error. The remaining helicopters arrived at Desert One, but when one of them malfunctioned Beckwith had no choice but to call the operation off, and Delta Force therefore loaded their equipment onto the Sea Stallions. The first heavily-laden helicopter took off, attempted to bank, stalled and crashed into one of the C-130 tankers. Both aircraft went up in flames and several servicemen died. After five hours at Desert One, the assault force finally withdrew – defeated by equipment failures outside their control.

A commission of enquiry under Admiral James Holloway pinned part of the blame on detailed tactical errors at the planning stage: more helicopters should have been made available, and insufficient attention was paid to the difficulties of flying helicopters across a wide expanse of desert.

Training had suffered because of over-zealous security, and Delta's troopers had not been provided with a mock-up of Desert One, even at the full dress rehearsal. However, the central difficulties appear to have been at the level of command and control. Unlike the British SAS, Delta had been completely dependent on other arms of the US military from both

a command and logistic point of view. The plan that was developed for Eagle Claw was unbelievably complex, involving contingents from the Rangers, Delta Force, USAF, the US Navy, the Marines, and agents of the Department of Defense acting in concert. As a result, there had been no overall commander on the ground.

Partly as a result of Eagle Claw's failure, Jimmy Carter lost the 1980 presidential elections, and in april 1981 Ronald Reagan took office. The new administration announced its determination to strengthen the Special Forces community and its intelligence gathering capabilities in response to international terrorism, and for Delta a new period of training was accompanied by a change in the command structure when Colonel Beckwith retired from the army in 1981. The US military establishment was quick to appreciate the lessons of failure at Desert One and in 1982 the 1st Special Operations Command (SOCOM) was created at Fort Bragg to coordinate the Special Operations Forces (SOF) of the army, air force and navy.

In July 1984, members of Delta Force were deployed as advisors when a Venezuelan DC-9 airliner was hijacked and diverted to the island of Curaçao in the Caribbean. A 12-man Venezuelan team successfully stormed the aircraft on 31 July in the presence of Delta operators, rescuing all 79 passengers and killing two of the terrorists. On 4 December, a Kuwaiti DC-9 airliner was hijacked by Shia Muslim militants on a flight from Kuwait to Karachi and forced to land at Tehran airport. Delta was deployed to Muscat in Oman and began rehearsing a rescue, but Iranian forces successfully stormed the plane.

On the morning of Friday, 14 June 1985, another group of Shia extremists hijacked a TWA Boeing 727 and took it to Beirut. Delta was deployed in readiness, apparently to Cyprus, but after a 17-day ordeal a negotiated release of the hostages was obtained. Five months later, Delta was once again deployed to the Mediterranean when an Egyptair 737 was hijacked and diverted to Valetta, Malta, during a flight from Cairo to Athens. However, Delta was denied an opportunity to storm the aircraft – with tragic results. Before Delta's arrival, late on 24 November, a team of Egyptian commandos stormed the plane and a fight

Left: An Iranian soldier surveys the carnage that marked the failure of Operation Eagle Claw. The dual problems of command and planning that plagued the mission have since been the subject of exhaustive enquiries undertaken by the joint chiefs of staff. Inter-service rivalry remains a thorn in the side of Delta Force, though recent developments may alleviate this problem. The US Congress has approved a plan to take control of the Special Forces community away from the individual service arms. It is intended to create a Special Operations Unified Command, with its own budget, procurements and planning staff. Below: The Egyptian anti-terrorist unit, deployed to Malta in November 1985, counts the cost of an operation in which 50 hostages lost their lives. Incidents of this kind have prompted a move towards the concept of 'forward deployment', with US negotiators asking West Germany, Britain and Jordan for forward bases to allow the rapid deployment of Delta Force to the Mediterranean and Middle East. Britain is reported to have agreed to this, and by September 1986 West Germany had agreed in principle to a 12-man Delta team being stationed on its territory.

ensued in which some 50 hostages lost their lives.

The difficulty of deploying Delta to Europe or the Middle East rapidly enough during a terrorist incident was highlighted on 5 September 1986 when hijackers boarded a Pan Am 747 at Karachi airport. Although Delta was given permission to land at Karachi, the force was unable to reach the scene of the incident in time. At about 2200, when the aircraft's lights failed due to lack of fuel in the generators, a force of Pakistani commandos stormed the 747. Some 20 passengers were killed in the bloodbath that followed.

Delta continues to co-operate closely with other Western counter-terrorist units

A major plank of the Reagan administration's anti-terrorist initiative is a US-led drive to increase co-operation between Western counter-terrorist forces, a drive that has caused some Pentagon officials to advocate the creation of a multi-national force in which, of course, Delta would have a leading role to play. Most counter-terrorist planners are against this, arguing that differing tactics and weapons would almost certainly lead to collective inefficiency. However, Delta continues to co-operate closely with other Western counter-terrorist units, and joint exercises with the SAS and with the West German GSG9 are held regularly.

Delta's role as the spearhead of US anti-terrorist forces may change as attempts to integrate the disparate counter-terrorist elements of the US armed forces continue. Delta's backup and counter-part units within the SOF include the army's Task Force 160 of the 101st Army Air Assault Division, which provides helicopter transport and attack support, and the navy's Sea-Air-Land (SEAL) Team 6 which has been allowed to develop an amphibious anti-terrorist role. Despite doubts that have occasionally been voiced about Delta's future, it seems certain that the Pentagon is committed to Delta Force, and that Delta operators will stay in the front line of the fight against terrorism for some time to come.

THE AUTHOR Barry Smith has taught politics at Exeter and Brunel Universities. He is a contributor to the journal *History of Political Thought* and has written a number of articles on military subjects.

MOROCCAN SPEARHEAD

Operating in the harsh terrain of the Italian Front, Moroccan mountain warriors, known as goumiers, earned a fearsome reputation in combat

WHEN GERMANY assumed control of northern France following the successful Blitzkrieg offensive of 1940, little effort was made to disband or even insist on a reduction of France's groups of colonial infantry. In Morocco, for example, there existed a substantial garrison of 57 goums (bodies of armed men), each consisting of between 175 and 200 trained Moroccan mountain tribesmen known as goumiers. Although at that time Germany did not have full control over French affairs, it was considered that those highly irregular irregulars were fit only for their prewar colonial policing duties, and thus would pose no threat in modern warfare.

However, in November 1942 the Allies landed in Northwest Africa (Operation Torch), meeting only token resistance from the Vichy French troops stationed there. The goums then threw in their lot with the British and American forces and, despite their being armed with only the most antiquated of weapons, they went on to play a significant role in the Tunisian campaign, especially during Rommel's offensive at Kasserine Pass and the Allied counteroffensive of early 1943, which culminated in the defeat of all the Axis forces in North Africa.

In the lull that preceded the invasion of Sicily and the Italian mainland, the French forces were provided with modern arms and equipment by the Americans, and underwent intensive training in their use. The resulting 'New Model' army was called the Corps Expeditionnaire Français (CEF) and overall command was given to General Alphonse Juin (later Marshal), himself Moroccan born. Eventually it would consist of eight divisions. Among the first to be reckoned ready for active service were the 2nd Moroccan Infantry Division (2 DIM) and the 4th Moroccan Mountain Division (4 DMM). Three tabors (battalions) of goumiers were included in 2 DIM, whose overall commander was General André Dody, while 4 DMM, under General François Sevez,

92

Below: French scout cars cross a pontoon bridge on the Fifth Army front. As part of the CEF, the goums formed the spearhead of an Allied offensive which breached the Gustav Line and then continued north, towards Florence. Left: As German resistance begins to crumble, goumiers advance towards Siena. Far left: A goumier hones the blade of his bayonet.

MOROCCAN GOUMS

In 1907 France sent an expeditionary force to Morocco to assist the sultan in his attempts to 'pacify' his deeply divided, turbulent country. By 1908 the force's commander, General d'Amade, was in need of extra men to maintain order in the areas he had subjugated, and his solution was to recruit small groups of mobile police from the local tribes. Each was known as a 'goum', the Moroccan term for 'a body of armed men', and they were so successful that in 1912 General Lyautey, the first governor general of the Moroccan Protectorate, prepared to standardise them for incorporation into the regular French Army. From then on, each goum comprised 147 infantrymen and 41 cavalry, commanded by two French officers, usually senior and junior lieutenants, and three NCOs. Then, to forestall the anti-colonial lobby in Paris, it was declared in 1914 that the goums were an integral part of the sultan's Armée du Maroc.

When World War I broke out, 14 goums were serving in the sultan's army. As fresh goums were raised, they were grouped in fours into new tactical units known as 'tabors', a term used by the 'tirailleur' regiments to denote a battalion. 'Pacification' campaigns began again within Morocco after the war, coming to an end in 1934. The number of goums in existence continued to grow, however, and in 1939 there were 57, forming a total of 14 tabors. Finally, just as the fall of France was imminent in 1940, plans were made to form 'groupements de tabors', each one to comprise three tabors.

had two tabors under General Augustin Guillaume attached.

The first contingent of the CEF to participate in the Italian campaign consisted of Dody's Moroccans and the 3rd Algerian Infantry Division (3 DIA). Assigned as part of General Mark Clark's Fifth US Army, it landed at Naples in late 1943. Juin was very anxious to get to grips with the enemy as soon as possible so as to be given the chance to vindicate the honour of French arms, so badly damaged by the disasters of 1940. To his dismay, however, it soon became evident that Clark intended to employ the French force in a purely supportive role.

The fight degenerated into a slogging match with grenades, bayonets and the goumiers' curved daggers

Nevertheless, it was not long before circumstances offered Juin a chance, and he seized it vigorously. Attempting to push into the mountainous hinterland of Naples, the Americans, particularly the 34th Division of Major-General John Lucas's VI US Corps, were halted by the determined resistance put up by the grenadiers of the German 305th Division. Juin quickly pointed out that the battlefield was ideally suited to the Moroccans, especially his hardy, mountain-bred goumiers. Clark, desperate for a breakthrough, ordered 2 DIM to take over in VI Corps' sector, if only temporarily.

On 16 December 1943 Dody launched his attack. In very poor conditions, his men went forward, the goums deployed on either wing, towards their first objective, the precipitous heights of Monte Pantanaro. General Juin noted:

'It was an anxious moment. The very reputation of the CEF depended that day on the failure or success of that first French division to go into action. But I had complete confidence in that splendid Moroccan unit, its commander, its officers and its men, recruited mostly from the

Middle Atlas mountains and thoroughly at home on a terrain such as that of the Abrizzis and its killing weather...'

As usual, the goums were deployed on the flanks, feeling the way ahead for the main body. But now, for the first time they were able to join in the final stage of the battle, fighting shoulder to shoulder with their country's 'tirailleurs' (riflemen) as the fight degenerated into a slogging match with grenades, bayonets, and the goumiers' curved daggers. It was tough going, but finally, on 21 December, the goumiers and tirailleurs drove their opponents from the peaks of Monte Pantanaro. The operation was completed on 26 December when tabors and the 8th Moroccan Tirailleurs stormed the last pocket of the German defence, the Mainarde Ridge, after yet another hand-to-hand encounter, this time with a Jäger battalion of the 5th Austrian Mountain Division, freshly arrived on the scene.

Clark was so impressed with this victory that he ordered the CEF to keep control of the whole of Lucas's sector. At the same time, Juin sent an urgent signal to General Henri Giraud, then supreme commander of all the French forces, requesting that the 4th Moroccan Mountain Division, together with General Guillaume's groups of tabors, be despatched immediately to reinforce the CEF before the next offensive, a request which was complied with promptly.

During the course of the next few months the goums gained considerable notoriety, not only with the enemy but also with their British and American allies. German prisoners of war confided to their interrogators that if it were known that goums were operating in their immediate sector they were liable to be gripped, if not by panic, then by a marked feeling of unease. The goumiers had quickly established for themselves a sinister reputation as skilled night-fighters, capable of penetrating solo, or in small groups, behind the lines to fall, without the slightest warning, from the rear on any lonely or

While a sherifian decree was signed in 1914, recognising the goums as a permanent element of the Moroccan sultan's Armée du Maroc, they were always fully dependent on the French government for their very existence, from their formation in 1908 until 1956, when Moroccan independence was followed by their incorporation into the Moroccan Army. The individual goumier received his pay, arms and ammunition from the French, and in return was required to fight for French interests.

Although supplied with French weaponry, which he supplemented with the Moroccan curved dagger, the goumier originally received no uniform, no rations, and no accommodation. On active service he wore his usual tribal dress, including a turban and a striped *djellabah* (which varied in colour according to his tribe). In earlier years, a *gandourah* (a form of loose cloak, invariably grey in colour) and *babooches* (heelless slippers) were worn, but during World War II the goumier was issued with such basic items as a steel helmet and boots.

Many aspects of the goumiers' origins were in evidence during World War II. Obliged to provide their own food, they were followed by herds of skinny goats ('rations on the hoof'), and their women-folk were in attendance to serve meals in their camps. Drawn from wild and warlike tribes, the behaviour of the goumiers off the battlefield could appal European observers. One Italian civilian claimed that 'the Moroccans flung themselves upon us like unchained demons,' and on occasion even hardened Allied servicemen were glad to part company with the ferocious hillmen of Morocco.

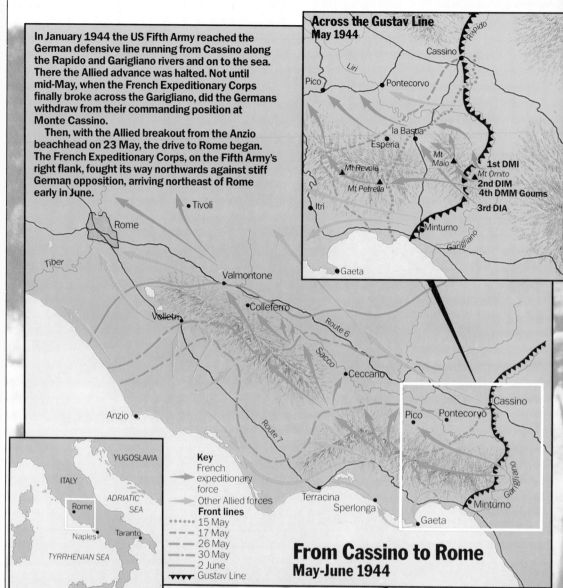

In January 1944 the US Fifth Army reached the German defensive line running from Cassino along the Rapido and Garigliano rivers and on to the sea. There the Allied advance was halted. Not until mid-May, when the French Expeditionary Corps finally broke across the Garigliano, did the Germans withdraw from their commanding position at Monte Cassino.

Then, with the Allied breakout from the Anzio beachhead on 23 May, the drive to Rome began. The French Expeditionary Corps, on the Fifth Army's right flank, fought its way northwards against stiff German opposition, arriving northeast of Rome early in June.

Across the Gustav Line
May 1944

Cassino
Pico
Liri
Pontecorvo
la Bastia
Esperia
Mt Revole
Mt Maio
Mt Petrella
Itri
Mt Ornito
1st DMI
2nd DIM
4th DMM Goums
3rd DIA
Minturno
Garigliano
Gaeta

Tivoli
Rome
Tiber
Valmontone
Colleferro
Route 6
Velletri
Sacco
Ceccano
Anzio
Route 7
Terracina
Sperlonga
Gaeta

Cassino
Pico
Pontecorvo
Garigliano
Minturno

YUGOSLAVIA
ITALY
Rome
ADRIATIC SEA
Naples
Taranto
TYRRHENIAN SEA
SICILY

Key
→ French expeditionary force
→ Other Allied forces
Front lines
⋯⋯ 15 May
– – 17 May
— — 26 May
–·–· 30 May
—— 2 June
▼▼▼ Gustav Line

From Cassino to Rome
May-June 1944

The goums were particularly well-suited to the rugged terrain of the Italian Front. Penetrating areas that were inaccessible to wheeled vehicles, these tireless North African warriors threaded their way through seemingly impassable mountain country to attack enemy lines. Left: The mastery of heavy firepower presented few problems for the versatile goumiers. Concealed in a hastily dug foxhole, a goumier ranges his mortar prior to bombarding an enemy strongpoint.

isolated position. Furthermore, it was known that the goumier was not at all interested in fighting a 'gentleman's war' or in taking prisoners. He was apt to measure the success of a mission by the number of heads he was able to take back to his CO.

One British officer remembered being quite amazed at the sight of a goum on the move. He likened it to the migration of one of the lost tribes of Israel, as they straggled along on either side of a hill track, clad in what he described as 'dirty striped blankets', their heads wrapped 'in equally dirty rags', while women and even some children trudged along in the rear, some leading goats and mules, the latter slung with bunches of trussed chickens.

American units of the US Fifth Army were warned that goums were in the neighbourhood and that should anything 'unusual' happen during the night they were to remain 'absolutely immobile'. One GI remembered that, while on duty in the early hours of one moonless night, he was 'shit scared' when he felt his helmet being gently raised while a hand, equally gently, began to massage the crown of his head. A goumier, it was explained later, was expert at 'recognising a Kraut scalp'.

Shortly after the arrival of the 4th Mountain Division and the tabor groups, the CEF staged its most successful operation of the whole campaign when it played a major part in the opening of the road to Rome. The general offensive which culminated in the capture of the Italian capital was launched simultaneously by the British Eighth Army on the right, the American Fifth Army on the left, and the CEF in the centre, spearheaded by the two Moroccan divisions and the tabors, with General Joseph de Monsabart's 3rd Algerian Division and a Free French Division, the 1st Division de Marche d'Infanterie (1 DMI), in support.

A gruelling task lay ahead. The first designated objectives were the well-defended Majo and Petrella massifs, which lay to the south of the Liri river and which were further protected by the formidable Gustav Line. Furthermore, German morale was high. Expecting the attack, General Raapke, commanding the 71st Division which was defending the Majo position, issued an Order of the Day: 'As at Cassino, the enemy will not pass. He will be stopped by our guns, our grenades and if necessary by our bayonets!'

H-hour was fixed for 2300 hours on 11 May 1944. Due to differences of opinion among the Allied commanders, preparations for the offensive had been unduly drawn out and the element of surprise had been lost. Thus, after a brief initial advance, the 4th and 8th Moroccan Tirailleurs were halted, while goumiers who managed to claw their way up the bared slopes of Monte Cerasole suddenly were stopped in their tracks by what was to them a new and terrifying weapon, the flame-thrower. Their retreat, in turn, exposed the flank of a supporting battalion of the DIM and those men also had to fall back, sustaining heavy casualties in the process.

When, at dawn on 12 May, Juin was informed of these early setbacks he was seriously worried. Having lived so much of his life in North Africa, he prided himself on his intimate knowledge of the Berber mentality, especially that of the goumier, who, he said, 'generally speaking is capable of a

Above: A conventional trim for an unconventional soldier. Exchanging his knife for a pair of scissors, a goumier cuts a compatriot's beard during a brief lull in the fighting.

long, but single, sustained effort, but it bodes ill if this effort is checked before it has attained its first objective.' Fortunately, in this instance his fears proved unjustified, for the withdrawals of the first night seemed only to create a fervent desire to get back at the enemy. The early morning of 12 May saw the offensive renewed with undiminished vigour as men of the mountain division and its goums hurled themselves against the Monte Revole and Monte Petrella strongpoints, while the 2nd Division swung to the right and completed the conquest of the Majo massif, thus opening the way for the advance on La Bastia and Esperian by the Algerians.

Juin's wisdom in demanding the presence of the Moroccan tabor groups and the mountain division was becoming increasingly evident. The country over which this battle for access to the plains was being fought was even more forbidding than that of the Moroccans' previous operations. The jumbled mass of jagged peaks and towering ridges presented a problem for even the most skilled mountaineer, and the precipitous slopes afforded not a scrap of cover. But as Juin had always expected, the goumiers were showing no sign of being daunted, either by the mountains or by the enemy.

At this stage of the campaign also, though losing nothing of their mobility, the goumiers were more heavily armed than ever before in their history. Many had become experts in the use of automatic weapons, such as the Bren machine gun and the Sten sub-machine gun. Individual goums could now also count on the additional firepower of two, sometimes three, mortars. These were normally transported on the backs of mules, but when the ground became too rough even for these sure-footed animals the bulky component parts and the supply of bombs were humped on the men's shoulders.

The ground over which the tabors resumed their assault on that morning was littered with their own dead and the shattered carcasses of their mules, but the initial shock of being confronted by flame-throwers had now passed. During the respite, they had rapidly devised a plan for

Above: The 2nd Moroccan Infantry Division moves by convoy into the mountainous hinterland around Naples. Deploying three tabors of goumiers on its flanks, the division struck terror into the heart of the enemy. Right: Equipped with steel helmets worn over their turbans, a goum machine-gun team pins down German positions in the ruins of a battle-torn village.

dealing with these frightening weapons. Having sited their mortars at close range, a small group would make a tentative thrust to tempt the enemy to open fire. Having thus located the enemy position, the mortar crews would promptly smother it with a shower of bombs. At the same time, every artillery piece fielded by the CEF was ordered to concentrate on the immediate targets of Monte Ornita, a dominant spur on the Majo massif, and Monte Petrella. This heavy bombardment was successful in causing a major disruption of the supplies and reinforcements that were being rushed to the enemy front.

By the afternoon of 13 May the tabors were steadily wiping out, one by one, the individual strongpoints installed on Monte Petrella, and General Dody's 2nd Moroccan Infantry Division finally succeeded in obliterating the remaining centres of resistance on the Majo massif, taking an abnormally high number of prisoners in the process. When the last German trooper in the sector had been either killed or disarmed, Dody ordered his prisoners to raise an enormous tricolor on the highest point of the massif, in order, he said, that, 'This emblem of French victory can be seen from the Tyrrhenian Sea to the Adriatic.'

On the morning of the 14th, all Juin's objectives had been taken, ahead of schedule, and the German front was effectively broken. The goumiers had overrun Monte Petrella and then, without a halt, pushed on to

crush the weakened and demoralised defences on Monte Revole.

On the right flank of the French force, however, the Eighth Army's desperate battle for Monte Cassino was still in progress, and the French slowed their advance to prevent their being cut off. When Monte Cassino fell on 18 May, however, the French advance was immediately resumed and on that same day the 4th Moroccan Mountain Division and the tabors reached the Pico-Itri road to link up with the left wing of the US II Corps.

By 30 May the Moroccans, now facing comparatively feeble resistance, had cleared the eastern slopes of the Lepini Range, while Dody's 2nd Moroccan Mountain Division, now strengthened by two more tabors, who were operating along the southern bank of the Sacco river, had successfully stormed Ceccino, Colleferro and finally Valmontone.

That period in Italy saw the finest achievements of the Moroccan fighting forces in World War II. In three weeks of continuous battle over stark mountains they had relentlessly pushed the Allied front forward against exceptionally stubborn resistance. In the process they suffered heavy losses, particularly among the goumiers who had fought in the forefront of battle throughout. With the Allies poised to debouche from the mountains to the northern plains, General Juin ordered the Moroccans to halt, while the 1st Division de Marche d'Infanterie and the Algerians moved into the van of the final advance on Rome.

THE AUTHOR The late Lieutenant-Colonel Patrick Turnbull commanded 'D' Force, Burma, during World War II. He published numerous books on a variety of military subjects following his retirement.

Left: Horses carry two fully laden goumiers as they pass through a small mountain village. Above left: Bastille Day, 1944, and a tribute to General Juin, commander of the CEF. Accompanied by General Clark of the US Fifth Army, Juin marches down an aisle flanked by standard bearers of the various French battalions.

After coming ashore at the Anzio beachhead, the 5th Battalion, The Grenadier Guards, fought one of the most furious defensive actions in the history of the regiment

'WHEN A GREAT DEAL is at stake, and failure or success depends upon a few men, war is raised from the level of mere operations to the level of drama.' These words appear in the regimental history of The Grenadier Guards, and they provide a special insight into why, of all the British units fighting at Anzio, the name of the 5th Battalion, Grenadier Guards, was given the distinction of being the first to be released by the censors to the world's press.

On 21 January 1944, the 5th Battalion was embarked in four landing craft heading out of the Bay of Naples. Together with the 1st Battalion, Scots Guards, and the 1st Battalion, Irish Guards, the Grenadiers comprised the 24th Guards Brigade, commanded by Brigadier A. Murray. The commander of the 5th Battalion, Lieutenant-Colonel G. C. Gordon-Lennox, had informed his men of their destination the previous night, and now he watched as the Allied invasion fleet of 243 vessels made its way out of the shadow of Mount Vesuvius. Part of a combined British and American assault force led by Major-General J. P. Lucas, the guardsmen were heading for Anzio, over 100 miles to the north. The air was brisk, but as the flotilla sailed past the island of Capri in the early morning sun, the sea sparkled like crushed sapphires.

The Factory was eventually taken after a nightmare hide-and-seek with German snipers

Air opposition failed to materialise, and by dawn on 22 January the 5th Battalion was only three miles off its appointed beach. Allied fighters circled overhead as the first troops went ashore, but enemy resistance was virtually non-existent. When the guardsmen arrived at the Anzio beachhead, there was no sign of any German counter-attack. An eerie silence descended, punctuated only by the sound of a few random shells exploding on the wet sand. By midnight, over 36,000 men and 3000 vehicles had been put ashore. The 5th Battalion waited patiently for its orders. They eventually arrived on the 23rd – the Grenadiers were to conduct a patrol to discover the location of the main German defences.

There was only one road that ran northwards from Anzio into the hinterland, and it was along this axis that Lieutenant J. Hargreaves' Grenadier patrol set out at dawn on 24 January. Travelling parallel to a railway track, the patrol's Bren-gun carriers and anti-tank guns arrived at a bridge known as the 'Flyover' without incident. The men were relaxed, and had even entertained the possibility of their advance taking them right up to the gates of Rome itself. A second bridge, known as the 'Embankment', was reached two miles further down the road. The town of Carroceto lay just ahead, and still no sign of the enemy.

Suddenly, shots rang out from a group of buildings several hundred yards ahead. The guardsmen knew that they had identified the first German road-block out of Anzio, and they withdrew to the Flyover. Upon receipt of this information, Brigadier Murray ordered the Guards Brigade to advance in strength

the following day. Once the enemy forces had be[en] cleared from Carroceto, the way would be clear fo[r a] break-out by the main Allied tank forces. T[he] German defenders had been forewarned, howeve[r] and a furious firefight erupted for control of a group [of] buildings known collectively as the 'Factory'. Hig[h] explosive and smoke shells were pumped into t[he] ranks of the advancing guardsmen as they rac[ed] across the open ground. The Factory was eventua[lly] taken after a nightmare hide-and-seek with Germ[an] snipers, but the fighting continued. Enemy forces remained entrenched in the surrounding area, and Grenadier casualties mounted steadily in the face of an intense artillery bombardment. The battalion had already lost 130 rank and file, and a wounded Gordon-Lennox had to be evacuated. He was later replaced as comman-

GUARDIANS OF ANZIO

ding officer by Lieutenant-Colonel A. Huntingdon.

The Factory became the lynchpin of the Allied advance into the hinterland, but intelligence now revealed that at least five German divisions were arrayed around the beachhead. Allied units continued to push forward in the hope of catching the enemy off-guard – but to little avail. Progress was painfully slow and, by 4 February, the advance of the 3rd Infantry Brigade towards Campoleone had ground to a halt. The guardsmen were forced to fight off a series of probing German counter-attacks, and it soon became clear that they would have difficulty in holding the ground they had gained. 'The situation,' wrote Brigadier Murray, 'was very serious.'

On the night of 6 February, the battalion's four companies were positioned two miles northwest of Carroceto, flanked by the Scots Guards on the right, and the Irish Guards to the left. The North Staffordshire Regiment lay to the south. The companies were perilously isolated from one another, with hundreds of yards of undefended ground lying between them. Listening to the shell-fire echoing all around them, the guardsmen knew that a fresh counter-attack was brewing – it was only a matter of time. Eight days of fighting had sapped the battalion's strength, but rest was impossible. Lieutenant P. Freyberg later wrote: 'We all had a blanket apiece, but that is not much when even the puddles freeze.'

During the evening of 7 February, a patrol from No. 1 Company spotted three columns of German infantry advancing on the Embankment from the west. Shortly after Lieutenant Hussey relayed this information to Lieutenant-Colonel Huntington at Battalion Headquarters, the enemy attacked simultaneously at several points on the Carroceto salient. The Grenadiers suddenly found themselves surrounded on all sides. Commanded by Captain N. Johnstone and Captain T. Browne respectively, No. 1 and No. 3 Companies fought a desperate

Main picture: Trenches such as this were turned into quagmires by the driving rain and sleet. Inset, far left: As a British soldier keeps watch for possible German snipers, piles of debris stand as testimony to the furious struggle for Carroceto. In the face of enemy fire, the guardsmen rushed towards a group of buildings known as the 'Factory' and secured their objective. The fighting continued, however, with German snipers taking refuge in isolated buildings (left).

THE GUARDS AT WAR

There were four lieutenant-colonels of The Grenadier Guards during the six years of war, and together they fostered the expansion of the regiment and sent six battalions overseas.

One of the most acute problems faced by the regiment was that of manpower. With large reserves in the Training and Holding Battalions, the 1st, 2nd and 3rd Battalions were raised relatively quickly at the beginning of World War II. Similarly, during 1941, the 4th, 5th and 6th Battalions were formed from an abundance of suitable recruits. However, when the battalions began to suffer heavy casualties in Europe and North Africa, the task of finding reinforcements of sufficient calibre became increasingly difficult. The height requirement for Grenadiers never fell below 5ft 9in throughout the war years, and the regiment was forced to compete with industry and the other services for manpower. Eventually, the regiment was unable to maintain the original number of battalions in the field – the 6th Battalion was withdrawn from active service in early 1944, and the 5th one year later.

Once a recruit had been accepted by the regiment, he would spend eight (later extended to 16) weeks at the Guards Depot, absorbing the essentials of drill and discipline that were the hallmarks of the Grenadiers on the battlefield. He then passed to the Training Battalion for a thorough grounding in field tactics and weapons training. Following this, each recruit was assigned – according to his ability and preferences – to one of the armoured or infantry battalions.

Above right: Guns at the ready, a patrol threads its way through the undergrowth in an attempt to discover the location of German troop deployments. Top right: The Ditch and the Gully – the scene of the 5th Battalion's memorable defensive action – can be seen in the bottom left of this aerial photograph.

delaying action until forced to fall back along the line of a railway that ran along the Embankment. Dodging from gully to gully, the 50 survivors found the whole area alive with German soldiers.

Meanwhile, Major W. Miller's No. 4 Company was under attack from the Buonriposo Ridge to the south. The North Staffordshires had been overrun and the air was thick with mortar shells, grenades and machine-gun fire. Miller attempted to withdraw to new positions just west of the gully that sheltered Brigade Headquarters, but both forward platoons

were overwhelmed by a remorseless tide of enemy infantry – but not without a fight. Lieutenant E. Collie took steady aim at a group of his adversaries and emptied his rifle and revolver into their ranks. A closing circle of German bayonets finally put an end to Collie's heroic stand.

Only No. 2 Company, commanded by Captain R. J. Martin, remained between the enemy and the men of Headquarters Company and Battalion Headquarters. All that was left of the battalion's defensive positions was a short, thin line that ended in a wide depression known as the 'Gully'. Beyond the Gully was the road to Anzio – if the Germans broke through, the Carroceto salient would be isolated and enemy tanks would have a clear route down to the Allied beachhead.

Having overwhelmed No. 4 Company by sheer weight of numbers, the German infantry massed on Buonriposo Ridge and sited machine guns in readiness for their attack on the Gully. Those guardsmen standing on the floor of the Gully never stood a chance and their ranks were decimated by a vicious stream of bullets. Headquarters Company returned the enemy fire with a vengeance. Lieutenant C. Hodson set up a 3in mortar and went to work, sending the rounds thudding into the midst of German troops. The battlefield was bathed in moonlight as the enemy, screaming and shouting, ran towards the guardsmen. The concentrated fire of Headquarters Company stemmed the tide for a few precious minutes but the German vanguard soon reached the

entrance to the Gully, having found a way over the 'Ditch', an overgrown irrigation channel that had impeded their advance.

At this critical moment, the true fighting spirit of the guardsmen asserted itself. The area opposite the Ditch was held by Major William Sidney, the commander of Support Company. Able to muster only a few men, Sidney ran to the edge of the Ditch and held his ground in the face of fierce enemy fire. Standing bolt upright, Sidney brandished his Thompson sub-machine gun and sprayed bullets down onto any German who dared show himself. When his gun jammed, Sidney withdrew and began to lob grenades at his adversaries. One of these exploded prematurely, wounding Sidney in the head and legs. Biting back the pain, Sidney refused to leave his post and continued to stem the enemy onslaught single-handed. When shrapnel from a German stick grenade finally forced him to retreat from the Ditch, Sidney was able to do so in the knowledge that the first enemy attack had been beaten, and that elements of Support Company had arrived as reinforcements. For his stalwart and heroic action, Major Sidney was awarded the Victoria Cross. On hearing of Sidney's award, a wounded guardsman praised the tenacity of his superior officer: 'Well, if he was as tough on the bloody Germans as he was on us, he deserves the VC.'

Back at the Ditch, however, the danger had not yet passed. Until the small party of Germans that remained entrenched on the other side of the rise could be dislodged, there was still a possibility that the enemy could launch a lightning counter-attack. Aware of this danger, Lieutenants G. Chaplin and W. Dugdale mustered a few guardsmen and scrambled up the ledge between the Gully and the Ditch. Silhouetted by the moonlight, the guardsmen started to work their way southwards, determined to seek out and crush German resistance in the immediate vicinity. Firing his Bren gun from the hip, Chaplin disposed of several Germans before his weapon jammed. Dugdale's sub-machine gun suffered a

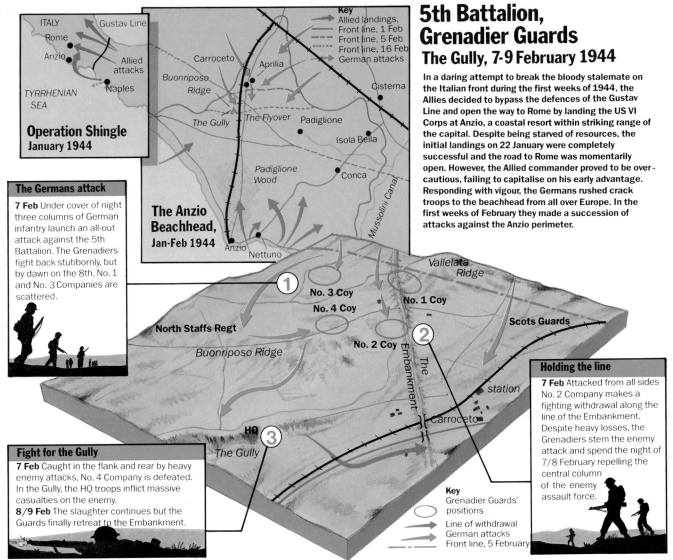

5th Battalion, Grenadier Guards
The Gully, 7-9 February 1944

In a daring attempt to break the bloody stalemate on the Italian front during the first weeks of 1944, the Allies decided to bypass the defences of the Gustav Line and open the way to Rome by landing the US VI Corps at Anzio, a coastal resort within striking range of the capital. Despite being starved of resources, the initial landings on 22 January were completely successful and the road to Rome was momentarily open. However, the Allied commander proved to be over-cautious, failing to capitalise on his early advantage. Responding with vigour, the Germans rushed crack troops to the beachhead from all over Europe. In the first weeks of February they made a succession of attacks against the Anzio perimeter.

Operation Shingle
January 1944

ITALY
Rome
Anzio
Gustav Line
Allied attacks
TYRRHENIAN SEA
Naples

Key
Allied landings,
Front line, 1 Feb
Front line, 5 Feb
Front line, 16 Feb
German attacks

Carroceto
Aprilia
Buonriposo Ridge
Cisterna
The Gully
The Flyover
Padiglione
Isola Bella
Padiglione Wood
Conca
Mussolini Canal

The Anzio Beachhead, Jan-Feb 1944
Anzio
Nettuno

The Germans attack

7 Feb Under cover of night three columns of German infantry launch an all-out attack against the 5th Battalion. The Grenadiers fight back stubbornly, but by dawn on the 8th, No. 1 and No. 3 Companies are scattered.

Vallelata Ridge
No. 3 Coy
No. 1 Coy
No. 4 Coy
North Staffs Regt
Scots Guards
Buonriposo Ridge
No. 2 Coy
The Embankment
station
Carroceto
HQ
The Gully

① ② ③

Holding the line

7 Feb Attacked from all sides No. 2 Company makes a fighting withdrawal along the line of the Embankment. Despite heavy losses, the Grenadiers stem the enemy attack and spend the night of 7/8 February repelling the central column of the enemy assault force.

Fight for the Gully

7 Feb Caught in the flank and rear by heavy enemy attacks, No. 4 Company is defeated. In the Gully, the HQ troops inflict massive casualties on the enemy.
8/9 Feb The slaughter continues but the Guards finally retreat to the Embankment.

Key
Grenadier Guards' positions
Line of withdrawal
German attacks
Front line, 5 February

similar fate. At this point, the guardsmen came face to face with a party of Germans bent on revenge. With tiger-like reflexes, Chaplin picked up a rifle from a fallen enemy soldier and returned their venomous fire. Realising that the odds were stacked against them, the men then jumped the 30ft down into the Gully – surviving the ordeal unscathed. The time was now 0330 hours, and a second German attack on the Gully was imminent.

Sensing that victory was still within their grasp, German soldiers launched themselves against the guardsmen. A constant volley of Grenadier small-arms fire, however, soon quelled the enemy's battle cries. They were further demoralised by the skill of the artillery officer, Major Greig, who directed the fire of his guns right onto the lips of the Gully itself. Daybreak was accompanied by a significant drop in the frequency of enemy attacks. Throughout the night the guardsmen had been reloading and firing without hesitation, and now the morning light afforded them the luxury of surveying the battlefield. Rivers of blood ran deep beneath the bodies of German soldiers that littered the ground, and the guardsmen were relieved to see the remaining elements of Captain Martin's No. 2 Company making their way along the Embankment. Isolated from Headquarters, the company had spent the night repelling the central German column.

The Gully was now held by 29 guardsmen and 45 Americans. While the latter were deployed to cover the track between the Gully and the Embankment, Huntington divided those of his men who were not wounded into three groups. Lieutenant Lyttleton and seven men were positioned to the south of the Gully; Lieutenants Dugdale and Hodson and seven more guardsmen were responsible for holding the northern entrance; and Lieutenant Chaplin, with eight men under his command, was ordered to cover the Ditch. The heavy rain that had been falling throughout the afternoon now turned to sleet, and the guardsmen braced themselves for the German attack.

The first assault was driven off, but the Germans were determined to break through. The Grenadiers' War Diary describes their doomed final attempt:
'There followed the now familiar advance to the

Although Brigadier Murray had realised that the area north of the Anzio beachhead was going to be difficult to hold against sustained German counter-attacks, he had no idea that the fate of the entire invasion force would depend on the 5th Battalion, Grenadier Guards. Below left: The battalion, with Major William Sidney in foreground, stands to attention during the presentation of battle honours. Below: General Alexander presents Sidney with the Victoria Cross in recognition of his bravery in the face of enemy fire. At the beginning of the war, those applying for commissions in the regiment had gone directly from civilian life to Sandhurst. If their performance as officer cadets was acceptable, they would be sent to the Training Battalion. Later, however, potential officers underwent the same discipline and training as a guardsman before passing to Sandhurst. Thus, through their own experience, Grenadier officers could gauge exactly what the men under their command were capable of in battle.

Ditch and the sight of more miserable Germans walking up and down trying to find a way across ... the enemy who were unwise enough to come within grenade-range were dealt with, and the stretch of the Ditch nearest to the defenders on the ledge was used as the killing ground.'

The men held their ground until the early hours of 9 February, but were forced to withdraw when the sleet turned the bed of the Gully into a muddy stream. Lieutenant-Colonel Huntington's men had done what had been asked of them and now, appreciating that they were weakened by their long ordeal, he organised a withdrawal through the southern exit of the Gully. After turning north and taking up new positions under the Embankment, the guardsmen wolfed down their first hot meal in 48 hours.

The battalion held its new position for a further 12 hours, despite the loss of Huntington, who fell victim to machine-gun fire during one of several fierce enemy attacks. The Scots Guards, mauled by German firepower, were forced to withdraw from Carroceto to the Embankment. The Allies were back where they had started; the element of surprise had vanished. However, for the 5th Battalion, Grenadier Guards, now held in reserve back at the Flyover, honour had been satisfied. On 12 February the commander of 15th Army Group, General Harold Alexander, paid tribute to the battalion, saying that they 'had made history and tradition worthy of the finest in all the military history of the Brigade of Guards.'

THE AUTHOR David Williams is a freelance writer who has contributed a number of articles to military publications, specialising in World War II.

AGONY OF THE LEGION

The French Foreign Legion units at Dien Bien Phu in 1954, doomed to be overwhelmed by the Viet Minh guerrillas, fought to the last man

Above: The agony of Dien Bien Phu begins. Legionnaires run for cover as the first salvoes of the Viet Minh attack hit home on 13 March, 1954. The enemy had unexpectedly concentrated large calibre guns around the base in the months prior to their assault.

FROM HIS headquarters, Lieutenant-Colonel Jules Gaucher, the commander of the central sector of the French base at Dien Bien Phu, could look northwest at the strongpoint called Gabrielle. Across the Nam Yum river, which meandered through the base, lay Anne-Marie and around Route 41, the river and the air strip stood Huguette, Claudine, Eliane and Dominique. With his field glasses he could pick out Isabelle, lying eight kilometres to the south. Gaucher

smiled to himself; like his own strongpoint, Beatrice these positions had been named after the mistresses of the base's commander, Colonel Christian de Castries.

According to the French generals in Hanoi, Dien Bien Phu was a natural fortress. Lying in an oak-lea shaped valley, the *cuvette* (bowl), as the base was known to the French, was protected from mass assaults by the 600m-high, scrub-covered hills that encircled its perimeter. Ridges, astride Route 41 and the river, gave the French natural defence lines which they had strengthened by building trenches and strongpoints. Gaucher, however, was less than happy. From the tactical point of view, he realised that the villages and hamlets in the valley were well within artillery range of the surrounding hills, and he knew that if General Vo Nguyen Giap, commander of the Viet Minh forces, could get his guns into these

CAMERONE

In 1892 the Legion raised a monument on the site of the little-known Battle of Camerone. Its simple inscription reads: 'Here stood fewer than 60 men against an entire army. Its weight overwhelmed them. Life, sooner than courage, forsook these soldiers of France.'

These few words sum up the valiant action fought in Mexico on 30 April 1863 by a column of Legionnaires under the command of a one-handed veteran of the Crimean War, Captain Danjou.

On that fateful day, these few men, ordered to meet a gold shipment travelling from Vera Cruz to Pueblo, faced the might of 2000 rebels who were fighting to overthrow the Emperor Maximilian.

At about 0530 hours, the column ran into an ambush and had to fight its way to the shelter of an abandoned farmhouse at Camerone. There was little chance of survival.

The uneven contest continued throughout the afternoon until only five men remained on their feet. Five men then charged 2000. Only three survived. Both they and the dead had fought in the highest tradition of the Legion, and their action is still commemorated to this day. Every year, Legionnaires assemble to acknowledge the valour of these men; an account of the battle is told to every recruit and at Aubagne, the Legion's headquarters, the wooden hand of Captain Danjou is paraded before the assembled crowd with great reverence.

In the years since Camerone, the Legion has lived up to the tradition and fighting spirit displayed by Danjou and his Legionnaires, and nowhere more so than at Dien Bien Phu.

positions, they would be in for a hell of a pounding. Unlike the French High Command, he was also aware that supplying the base would be difficult. The valley had double the average rainfall of Indochina and a mist, known as the *crachin*, made flying dangerous for much of the year.

Gaucher had spoken of his fears to Colonel Pierre Langlais, the base's second-in-command, and had been less than impressed by Colonel Charles Piroth, the artillery commander, who believed that his guns could blast the Viet Minh out of the hills if called upon to do so.

Gaucher had seen the base grow since late 1953 when Indo-Chinese of the 6e Bataillon de Parachutiste Coloniale (6 BPC – 6th Colonial Parachute Battalion) under Major Marcel Bigeard, and Legionnaires of the 2nd Battalion, 1er Régiment de Chasseurs Parachutistes (1 RCP – 1st Parachute Light Infantry Regiment) and his own unit the 13e Demi-Brigade de la Légion Etrangère (13 DBLE – 13th Foreign Legion Half-Brigade) had extended the airstrip and built the first strongpoints. Over the following months, he had seen reinforcements arrive until the garrison amounted to 10,814 men. Of these, 2969 Legionnaires from the 13 DBLE, the 2e and 3e Régiments Etranger d'Infanterie (2 and 3 REI – 2nd and 3rd Foreign Legion Infantry Regiments) – and the 1er Bataillon Etranger de Parachutistes (1 BEP – 1st Foreign Legion Parachute Battalion) formed the core of the defenders' strength. Artillery support was provided by 24 105mm and four 155mm howitzers, three 120mm mortar companies and a squadron of 10 M-24 Chaffee light tanks.

By March 1954 it was clear that Giap intended to storm Dien Bien Phu. Over the preceding months he had concentrated the cream of his army, 28 infantry battalions with a strength of 37,500 men, from the 304th, 308th, 312th and 316th Infantry Divisions, together with the 351st Heavy Division, around the base.

The Viet Minh attack on Dien Bien Phu began at 1700 hours on 13 March. The effectiveness of the initial bombardment against the outlying strongpoints stunned the French and set the scene for their eventual defeat. Sergeant Kubiak of the 3rd Battalion, 13 DBLE felt the fury of the barrage:

'It was about 1700 hours and I was walking back to my pillbox. At that very moment, without warning, I felt that the end of the world had come. The whole of the strongpoint (Beatrice) went up in smoke. All around me the ground erupted. I saw men falling and lying still. Shells were raining down like hailstones in a winter storm. Pillbox after pillbox and trench after trench was being blown to pieces. How the hell had the Viets managed to bring up the guns needed for such a barrage?'

The tremendous bombardment reduced many of Beatrice's fixed defences to a mass of rubble and caused heavy casualties among the Legionnaires. Their commander, Major Paul Pégot, was one of the first

to be killed. When informed of his death, Gaucher called for his second-in-command Major Michel Vadot, who was in an adjoining bunker, to join him. 'We must send someone to take Pégot's place immediately,' he said, as Vadot arrived. 'I suggest…' At that very moment a 105mm shell smashed through the bunker's flimsy roof and exploded. A Legionnaire who was on the scene moments later, remembered the carnage caused by the shell:

'The colonel was lying in the debris of his desk, his legs smashed, his chest torn open, but still alive. Beside him lay his aides, Lieutenants Bailly and Betteville, together with the radio operator, killed outright. Only Vadot, his chest peppered with splinters, was capable of carrying on.'

For more than eight hours, the Legionnaires holding Beatrice hung on to their shattered positions in the face of repeated suicidal assaults by the Viet Minh. Their dead were piled so high in front of the trenches that they formed a shield that protected the following waves of attackers. Kubiak was one of the Legionnaires who had to face the onslaught:

'The Viet barrage lifted and their infantry were through the second hedge of barbed wire. It was then that I gave the order to open fire. The trouble was that for every one that went down there were two to take his place. An army of ants! We must have gone on firing for hours. The dead and dying piled up but, in spite of the gaps we tore through their ranks, they kept coming on.

'Suddenly, I realised that it couldn't be long before they broke right into what was left of our pillbox. This had to be prevented. If that had happened there would be at least 10 of them to every one of us. It was only then that I was aware that night had fallen. We kept firing and by some miracle managed to stop those human ants.'

By 0300 hours (14 March), it was clear that de Castries did not intend to put in a counter-attack and that, to survive, the Legionnaires in Beatrice would have to fight their way back to the central sector.

Legionnaire, French Foreign Legion, Dien Bien Phu 1954

Even as late as 1954 French troops were still wearing uniforms made from materials provided from British and American stocks of World War II vintage. This soldier wears British 'windproof' clothing, re-cut to French Army pattern. Webbing and armament is of US origin, notably the 0.3in M2 carbine with folding stock for airborne troops. This lightweight weapon was widely distributed to French troops in Indochina, even though, with its small cartridge, it lacked the power and range of the conventional rifle. The M2 version was a development of the M1A1 carbine, fitted with a fire-selection feature which allowed it to be used as a sub-machine gun. The rapid firepower of the M2 was an effective counter to the human-wave attacks of the Viet Minh mounted against the French positions at Dien Bien Phu. Using either a 15 or a 30-round box magazine the M2 had a maximum range of 250m.

Above: Legionnaires leading a counter-attack across the shell-blasted battlefield of Dien Bien Phu in a desperate attempt to reach their comrades trapped in Gabrielle. Right: The French faced a totally unexpected volume of fire that ripped through their ranks and many, like this man, were hit before they could get to grips with the enemy. Below right: A French mortar crew in action.

As Colonel Gaucher had died within half an hour of being hit, command was passed to Vadot. Known to his men as 'Papa', he was admired for his calmness under fire. Although in great pain, he coolly picked up the bunker's field telephone, which was still functioning somehow, and called the command post. 'Vadot here,' he said. 'We will hold to the last, but that cannot be for long. Request maximum fire on our own positions.' De Castries' reaction was to order the immediate withdrawal of the remnants of 13 DBLE.

The wire in front of their position was covered with dead and dying

The following evening (15 March) it was the turn of Gabrielle, held by Algerians and a Legion mortar unit commanded by Lieutenant Erwan Bergot, to face the ferocious might of the Viet Minh. Although their attacks lacked the fury of those launched against Beatrice, only one sector of the bastion remained in French hands by dawn. Bergot remembered the devastation wrought on the defenders during that bloody night:

'Morning came. In the weak light of a hesitant dawn, I remember Roll coming to sit on the edge of my dug-out. He had grey hair. "We've aged 10 years," he said with a grin. Day broke. We thought we had been at the centre of the cataclysm, but we quickly found that things were the same all over. Everything was smoke-blackened, broken, smashed, demolished.

'All around, wandering silhouettes bore wit-

PREPARING FOR DISASTER

In February 1950, four years after he had been forced to evacuate Hanoi in the face of a French expeditionary force, General Vo Nguyen Giap, military commander of the Viet Minh, felt confident enough to unleash the army he had been building up against the scattered French garrisons of northern Tonkin. In a series of small-scale offensives which culminated in the attack on the Cao Bang ridge in late 1950, Giap forced the French to relinquish control of the area and retreat to the comparative safety of the Red River delta.

With his supply routes to China secure, Giap was able to concentrate his forces for the proposed destruction of the French defences. The French, however, were stronger and better prepared than the Cao Bang debacle had suggested and the Viet Minh suffered heavy losses in the delta at Vinh Yen, Mao Khe and Phat Diem between January and June 1951.

The victories of 1951 inspired the French to attempt a break out from their positions in the delta – at Hoa Binh in November 1951, and against the Viet Minh supply dumps at Phu Tho and Phu Doan (Operation Lorraine: October/November 1952). Both offensives ended with the over-extended French forces withdrawing behind their defences once again, and the capture of the Nghia Lo Ridge by the Viet Minh in October 1952 only served to confirm to the French that they had lost the initiative in the war.

On 9 April 1953 Giap launched his troops against Laos with the twin aims of joining forces with local guerrillas and seizing the country's valuable opium crop. The invasion was, however, blunted by the French, whose men, deployed in a series of 'points of resistance' along the Laotian border, were able to smash wave after wave of the enemy.

Although it was abundantly clear that Giap had suffered a major setback, the French made the grave mistake of misinterpreting events. Victory led the High Command to believe that they had a firepower that was immeasureably superior to the Viet Minh's; that Giap would be unwilling to fight a major battle for some time and that centres of resistance, supplied from the air could hold out against Viet Minh assaults indefinitely.

It is only within this context that the decision to establish a base at Dien Bien Phu can be understood. On 25 July 1953, the French generals in Hanoi agreed to the creation of a base in the area as part of a preventative action against future Viet Minh drives into northern Laos. Their plan, codenamed, Operation Castor, began on 20 November 1953, when paratroopers landed at drop zone 'Natasha', a few hundred metres from the village of Dien Bien Phu.

Five hours after the first landings, Natasha was secure enough to permit the dropping of the remainder of the task force. The first phase of the operation had undoubtedly gone well. Opposition was minimal: the French secured the area at a cost of only 63 casualties, and had killed at least 50 Viet Minh.

Below: A few days after the opening of the most intense barrage yet seen in Indochina, the French defences at Dien Bien Phu were shattered beyond repair. The men in the trenches paid the price for their generals inability to recognise that the Viet Minh were capable of concentrating their artillery in the hills around the base. Many bunkers and trenches were covered with little more than flimsy wooden planks and a layer of earth. The strain of almost continuous bombardment threatened to undermine the French will to resist and it was only the fortitude of the Legion that enabled them to hold the base for 57 days.

ness to the disaster. They were digging in the ground, uncovering corpses, carrying the wounded without a word, uttering no cry, deaf and blind to their surroundings.'

The surviving Legionnaires had little time to collect their thoughts before the Viet Minh launched another attack. Legionnaires Zimmerman and Pusch faced the main Viet Minh thrust. Both had armed themselves with automatic rifles and then had occupied a trench from where they were able to beat off repeated enemy assaults. So determined was their defence, that the barbed wire in front of their position was covered with dead and dying. On the verge of being overrun, they threw eight grenades in quick succession and then hacked their way back to the main body of the defenders.

The fate of Gabrielle was clearly in the balance and, although de Castries decided to throw in a counter-attack to relieve the pressure, he made the mistake of not committing the whole of the force at his disposal. Only two companies of Legion paras, a battalion of Indochinese and a troop of Chaffee tanks were sent into battle.

Only the Legionnaires pressed forward to help the garrison of Gabrielle repulse the Viet Minh

On reaching the ford of the Nam Yum, the column came under heavy fire. It was too much for the Indochinese, who broke and scattered. One of the tanks was hit and the rest turned back. Only the Legionnaires pressed forward to help the garrison of Gabrielle to repulse the Viet Minh. Such a small force, however, had little chance of survival and, that afternoon, they were ordered to retreat.

Giap had won the strategic ridges to the north and northeast of Dien Bien Phu at the sacrifice of 3000 men. With his artillery and mortars strung out along the surrounding heights he could crucify the French troops in the valley below. The long, slow agony of Dien Bien Phu had begun and even the lowliest

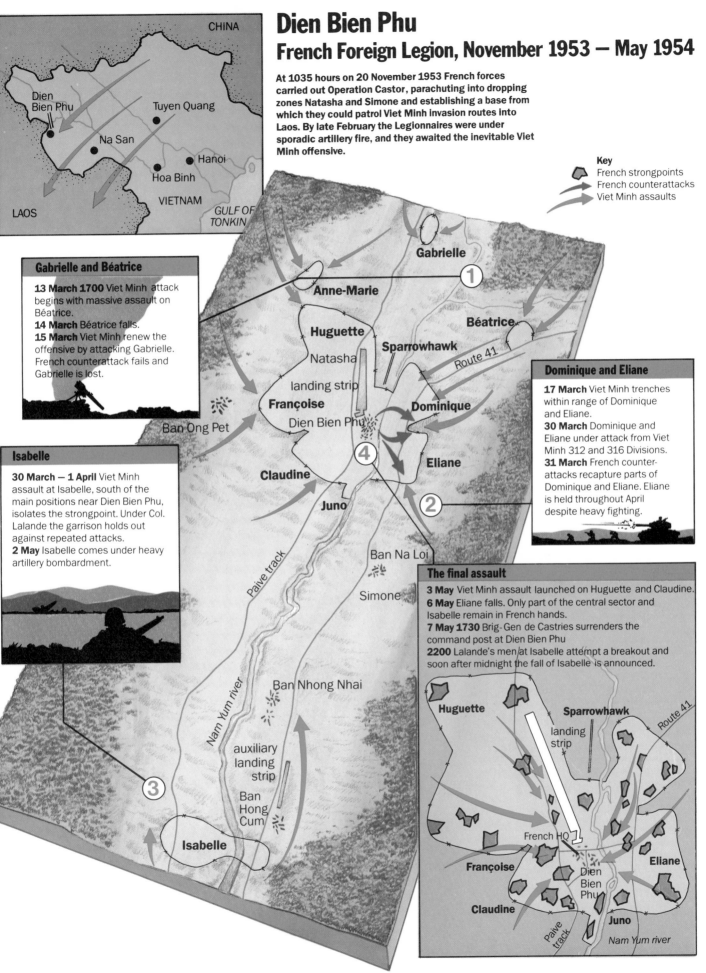

Dien Bien Phu
French Foreign Legion, November 1953 — May 1954

At 1035 hours on 20 November 1953 French forces carried out Operation Castor, parachuting into dropping zones Natasha and Simone and establishing a base from which they could patrol Viet Minh invasion routes into Laos. By late February the Legionnaires were under sporadic artillery fire, and they awaited the inevitable Viet Minh offensive.

CHINA

Dien Bien Phu

Tuyen Quang

Na San

Hanoi

Hoa Binh

VIETNAM

LAOS

GULF OF TONKIN

Key
French strongpoints
French counterattacks
Viet Minh assaults

Gabrielle and Béatrice

13 March 1700 Viet Minh attack begins with massive assault on Béatrice.
14 March Béatrice falls.
15 March Viet Minh renew the offensive by attacking Gabrielle. French counterattack fails and Gabrielle is lost.

Isabelle

30 March — 1 April Viet Minh assault at Isabelle, south of the main positions near Dien Bien Phu, isolates the strongpoint. Under Col. Lalande the garrison holds out against repeated attacks.
2 May Isabelle comes under heavy artillery bombardment.

Dominique and Eliane

17 March Viet Minh trenches within range of Dominique and Eliane.
30 March Dominique and Eliane under attack from Viet Minh 312 and 316 Divisions.
31 March French counter-attacks recapture parts of Dominique and Eliane. Eliane is held throughout April despite heavy fighting.

The final assault

3 May Viet Minh assault launched on Huguette and Claudine.
6 May Eliane falls. Only part of the central sector and Isabelle remain in French hands.
7 May 1730 Brig- Gen de Castries surrenders the command post at Dien Bien Phu
2200 Lalande's men at Isabelle attempt a breakout and soon after midnight the fall of Isabelle is announced.

Gabrielle

Anne-Marie

Béatrice

Huguette

Sparrowhawk

Route 41

Natasha

landing strip

Françoise

Dien Bien Phu

Dominique

Claudine

Eliane

Juno

Ban Ong Pet

Ban Na Loi

Simone

Paive track

Nam Yum river

Ban Nhong Nhai

auxiliary landing strip

Ban Hong Cum

Isabelle

Huguette

Sparrowhawk

Route 41

landing strip

French HQ

Françoise

Dien Bien Phu

Eliane

Claudine

Juno

Paive track

Nam Yum river

① ② ③ ④

THE VIET MINH

By 1941 the communist-dominated Viet Minh had become the driving force behind a nationalist movement which had been in direct conflict with the French since the early 1920s. During 1945, the Japanese took control of Indochina and the Viet Minh redirected their small-scale operations against their new masters.

The defeat of the Japanese, however, did not herald independence and the return of the French in 1946 forced an escalation in the conflict.

Under the command of General Vo Nguyen Giap, the Viet Minh were expanded and reorganised into three main elements. At the most basic level were the militia; poorly armed and with little training, they formed the backbone of the army and by 1954 the militia fielded over 350,000.

On the second level of Giap's army were the regional troops. Better trained and equipped than the militia, they provided intelligence and made good use of their local knowledge to harass the French forces.

Their strength was estimated to be 75,000 at the time of the French defeat in 1954.

The top level of Giap's new military structure was the regular force, or Chu Luc. Built up in the Viet Minh's northern bases during the 1940s, it was ready for action from 1950 onwards. During 1950, its 60 battalions were allocated to five divisions. Each division was conventionally organised with three infantry regiments and support weapons. By the time of Dien Bien Phu, Giap had 125,000 regular troops at his command.

Although the Chu Luc had been carefully built up, they were used in massed infantry attacks that were often wasteful, but Giap knew he could replace his losses and his ability to maintain the initiative eventually forced the French to concede defeat.

coolie in the camp recognised that the battle could go only one way. Colonel Charles Piroth, the one-armed artillery commander, blamed himself for not having silenced the guerrillas' artillery. 'It's finished,' he wept to Langlais. 'We're going to be massacred and it's all my fault.' He entered his dugout and, being unable to cock a pistol with his single hand, he primed a grenade and blew himself to pieces.

On 17 March the T'ai battalion holding Anne-Marie deserted en masse, leaving the only remaining bastion in the north, along with their weapons and equipment, to the enemy. By now, more than 40,000 Viet Minh surrounded Dien Bien Phu and, as Giap's gunners shelled the runway and the nerve centre of the camp round the clock, his sappers cut trenches to bring the infantry within rifle range of Huguette, Claudine, Dominique, Eliane and Isabelle. With the airstrip under intense bombardment, any thoughts of wholesale evacuation evaporated. Even the wounded, crowded into the camp's medical post, knew that there would be no escape.

The defenders had to face the daunting fact that the airstrip, openly referred to as Dien Bien Phu's 'lifeline', could no longer be used during the day; by the 23rd, the Viet Minh artillery was cratering the runway so effectively that even night flights had to be abandoned. For the rest of the siege, the garrison would have to rely on parachute drops during the hours of darkness for resupply and reinforcement.

As a result of their suicidal attacks, the Viet Minh had also suffered heavy losses and, during the closing days of March, Giap acted against the advice of his Chinese aides and ordered the remaining French positions to be reduced by artillery fire. The lull, however, was only temporary. The beginning of April saw the first of several attempts by the Viet Minh to capture the main bastions simultaneously.

The focus for the new attacks was Isabelle, the only remaining strongpoint outside the central perimeter. The position, however, was a tough nut to crack. Its commander, Colonel André Lalande, had a strong garrison consisting of 3/3 REI, a battalion of

Algerians, a Moroccan unit, a battery of 105mm guns and a troop of Chaffee tanks. Lying over five kilometres from the central sector, Isabelle was soon isolated but, with so powerful a garrison, repeated assaults were beaten off throughout the siege.

At this crucial stage, the French High Command ordered two fresh parachute battalions to be dropped into the *cuvette*. The first to arrive on 11/12 April was the 2e Bataillon Etrangèr Parachutiste (2 BEP – 2nd Foreign Legion Parachute Battalion). Conditions could hardly have been worse. Heavy rainfall and poor visibility forced the troop-carrying Daks (DC-3 Dakotas) to fly very low for the approach run. Incredibly, very few were hit by the barrage of anti-aircraft fire thrown up by the Viet Minh, but the Legionnaires suffered severe losses as they parachuted in. Many were hit while still in the air, some fell on to enemy positions and were bayonetted before they could disentangle themselves from their parachutes and others landed in minefields. One of the platoons that did reach the ground intact was thrown into the battle raging around Huguette. Within the hour, its two officers had been killed and only seven Legionnaires remained on their feet.

The Legionnaires were decimated when they charged at the enemy

On resuming their offensive after being rebuffed by Lalande's men, the Viet Minh concentrated on Huguette, with diversionary attacks against Eliane and Claudine. Unlike the outer defences, these strongpoints did not consist of one blockhouse, but a series of inter-connecting posts enclosed by a wire perimeter. These posts were known as *Points d'Appui* (PA) and were designated by numbers after the initial letter of the main strongpoint. In the battle for Huguette 1 (H1) on 23 April, the Legionnaires were decimated when they charged at the enemy with the bayonet. An eyewitness remembered the volume of

fire they faced:

'Enemy gunners crawled from their dugouts and, supported by the remaining Viet Minh machine-gun nests in no-man's land north of H2, began to lay down an almost impenetrable defensive fire in the best traditions of World War I.'

The equivalent of two companies was destroyed within minutes and, as a result of the grave losses they had suffered, the two para battalions were reformed into a Bataillon de Marche under the command of Major Giraud of 1 BEP. Its total strength barely amounted to that of two normal-strength companies.

On the morning of 30 April, the anniversary of Camerone, Captain Coutant, the commander of the 1st Battalion, 13 DBLE, at Eliane, was told that a consignment of *vinogel*, described as 'a peculiar purplish, jellified substance with a high alcohol content', had been dropped by parachute, but had landed behind enemy lines. At first, Coutant refused his men permission to retrieve this precious cargo, but by 2200 hours he had relented and had given a six-man squad leave to recover the 'wine'. One of the the Legionnaires present saw the recovery:

'At a given signal the Legionnaires dashed forward, crept along the wire, and then took cover behind heaps of debris and corpses. Fifty metres is not a long way under normal circumstances, but it's a very long way in a no-man's-land swept by the fire from hundreds of rifles.

'A whistle rang out! The three men who formed the diversionary group went into action; hurling grenades through the slits of the Viet bunkers. Meanwhile, the *vinogel* was passed by hand back to our lines.'

As the men drank their cups of *vinogel*, a voice called to them from the enemy positions. 'Why continue to fight?' it said. 'Why do what the Legionnaires at Camerone did – get yourselves massacred?' The French silenced this impudence by singing their marching song, *Le Boudin*.

By the following morning de Castries realised that the battle was lost. He had too few men to defend the battered perimeter. Many of the North Africans had lost the will to fight and had deserted their posts to take refuge in the caves in the banks of the Nam Yum. The burden of the defence had now fallen on the shoulders of the Legion, and the colonial paras, but they could not hope to succeed without ammunition and food. De Castries' men were too weak from exposure and poor feeding to carry on for much longer and reports of unwounded troops falling dead at their posts from sheer exhaustion only served to confirm their commander's worst fears.

On 3 May the Viet Minh redoubled their attacks on Claudine and Huguette. One by one the PAs were overrun. Kubiak, one of the men defending these strongpoints, recalled the desperate struggle:

'It was all hell let loose and the Legionnaires went berserk. And it didn't end with the dawn. Time after time the positions changed hands. We were all dead-beat. On our feet by some miracle. Obeying as if we were robots.'

Captain Philippe of 13 DBLE ordered one of his platoons to help in the defence of Claudine 5 (C5) where the Legionnaires were particularly hard pressed. In the absence of artillery support the men were soon pinned down. Philippe led the sappers of the 1st Battalion, 2 REI forward. A Legionnaire witnessed the relief force's attack:

'Then the sappers of 2 REI pushed forward. Placing themselves at the head of the company and Captain Philippe's platoon, they charged. The sight of these bare-chested and bearded giants was too much for the Viet Minh assault troops. By 2230 hours the Legionnaires were back inside the featureless hollow, filled with craters and bodies, that had been C5. Finally, these men were forced out and C5 fell for good at 0200 on 6 May. The Legionnaires were either dead or wounded.'

Now only Eliane stood between Giap and final victory. For three days the Legionnaires defending this battered strongpoint had weathered everything the Viet Minh could throw at them – shells, grenades, artillery and, something new, 'Stalin Organs', rocket batteries from Russia. If Eliane fell, the final hope of

Left: The architect of the Viet Minh victory, General Giap (standing), confers with Ho Chi Minh, (centre, sitting). Above: Viet Minh positioning an artillery piece. Right: Legionnaires mount a counter attack.

DISSENT AT THE TOP

The overall responsibility for the defence of Dien Bien Phu rested with a distinguished army commander, Colonel Christian de Castries. A cavalryman from a military family, he was an enthusiastic exponent of the aggressive approach that was the rationale behind the creation of the camp at Dien Bien Phu.

In consequence, it seems he did little to provide the camp with adequate protection against artillery fire and, when massed Viet Minh assaults began to overwhelm his outposts, he appeared to be totally unprepared to deal with the unexpected crisis.

In retrospect, he may have been the wrong choice for the appointment. His attitudes failed to impress the hard-bitten officers of the Legion parachute battalions at Dien Bien Phu. Socially, they were of a different class – Major Marcel Bigeard of 6 BCP, for example, was the son of a railwayman – and the idea of an 'honourable defeat' was anathema to them.

After the fall of strongpoints Beatrice and Gabrielle, Bigeard and Lieutenant-Colonel Pierre Langlais, a fellow paratrooper officer, sensed that with de Castries at the helm, victory would be impossible.

On 24 March, matters came to a head. Langlais confronted de Castries in his command post and stated that henceforth he and other Legion officers would direct the defence of the base. De Castries, still shocked by the deaths of several of his aides, accepted this ultimatum without argument.

The offensive spirit of the French may have been rekindled by Langlais' move but, by this late stage, the death knell of Dien Bien Phu had already been rung.

saving Dien Bien Phu and the 2000 men who could still carry a weapon would die.

On 6 May the French flag flew over only a fragment of the central sector and Isabelle. Only three French guns could reply to the Viet Minh's 400 that heralded the death knell of Dien Bien Phu. Giap's 'death volunteers' had hurled the remaining Legionnaires off Eliane and across the Nam Yum into Claudine. There was no let up for the French; the Viet Minh assault groups edged their way towards de Castries' command post and their artillery pounded what remained of the trenches. By mid-morning of the 7th, the Viet Minh were within 400m of victory. The paratroopers of the Bataillon de Marche made one last effort to stave off defeat. Kubiak was one of the Legionnaires present:

'Suddenly, I heard a shout, "Look, over there, to the left." I doubled across and saw about 100 Viets splashing across the Nam Yum. Before I could take any action some of the wounded had opened up. The firing was heaviest from one of our old pillboxes and Viets were dropping. I sprinted over to it and could hardly believe my eyes. It was a wounded man who was firing the machine gun,

INTO CAPTIVITY

Sergeant Kubiak of the 3rd Battalion, 13 DBLE, was the only survivor of Dien Bien Phu to record his impressions of the 57-day siege in a diary. This extract concerns his capture by the Viet Minh after de Castries had surrendered the base on 7 May:

'At that moment the Viets started crossing the river, with the idea of taking us prisoner I suppose. Then the balloon really went up. The wounded had no intention of surrendering and it was they who started up again. The Viets didn't just wait to be hit. After the first shot they concentrated all they had got on the few square metres we were holding. I can remember those last few moments very well indeed; a shell exploded under my nose and down I went, into the trench bottom's mud.

'There were shouts, yells and death rattles as the Viets swarmed in and finished off the wounded. A Viet spotted me and I guessed I'd had it. He raised his rifle, taking deliberate aim as he approached. I gritted my teeth, quite unable to move. I watched his finger press the trigger but, as the shot went off, he pitched forward and fell on top of me. As his body landed on me, everything went blank.

'I learned that he was shot by a Legionnaire and that his body had saved me from his murderous friends. The battle was over. My life as a prisoner had begun.'

The French commander at Dien Bien Phu, Colonel Christian de Castries (left), found it difficult to cope with the pressures of the battle and on 24 March was, in effect, replaced by a group of para and Legion officers led by Lieutenant-Colonel Pierre Langlais (far left). Their take-over, however, only delayed the inevitable and by 7 May the French, having lost the means to continue the unequal struggle, capitulated. Below left: Many Legionnaires, although suffering from shell-shock, remained willing to carry on the fight and to many death was preferable to surrender. Below right: Few of the 7000 French troops who were marched into captivity survived their ordeal.

with another wounded man acting as loader. "You see, Sergeant," said the gunner grinning at me, "I may have a few holes in my guts, but I can still fight. Don't leave me behind."

'We pushed the Viets back and then, without warning, their artillery stopped firing. I heard our commanding officer swearing like a navvy. I looked in the direction he was staring, and what did I see? There, over the general's command post, floated a huge white flag. The French guns had fired their last shells, the infantry were down to their last rounds, their last grenades and their rations were almost finished.'

In the final hours of the battle, the French High Command had suggested to de Castries that he should try to break out of Dien Bien Phu, but he had rejected the plan: 'Any breakout is bound to fail. We must cease firing.' The battle ended at 1700 hours. Only Lalande at Isabelle was in a position to execute 'Albatross', the proposed withdrawal plan. At midnight, after having split the garrison into three groups, he gave the order to move. All three groups penetrated the Viet Minh positions around Isabelle's perimeter, but only 12 men reached safety and the

rest, including Lalande, were taken prisoner and returned to Dien Bien Phu, from whence began the long march to captivity. Behind them lay 4000 dead, including 1500 Legionnaires. Over 7000 men set off into the jungle, including 4000 Legionnaires, but few lived to see their native shores again. The Viet Minh had lost 8000 dead and 15,000 wounded.

For 57 days over 5000 Legionnaires had faced the full might of the Viet Minh army and bombardment by hundreds of heavy guns. Those who died and the few that survived, had fought to the last in the highest traditions of the Foreign Legion. Their defeat did crippling damage to the French government's will to continue the struggle, and, later that year they signed a treaty with the Viet Minh that effectively ended their involvement in Indochina. The Legion was forced to quit Tonkin for good and only the graves of the dead would remain as a testament to their bravery. Between 1935 and 1954, 30,000 Legionnaires had been deployed in Indochina and some 309 officers, 1082 NCOs and 9092 Legionnaires had fallen in the service of their adopted country.

THE AUTHOR Lieutenant-Colonel Patrick Turnbull commanded 'D' Force, Burma, during World War II. He has published numerous books including *The Foreign Legion*.

GREY'S SCOUTS

At the time of the war in Rhodesia, the concept of mounted infantry was not new to southern Africa. The idea had worked well during the days of the Boer Commandos and South African Frontier Wars, and can be traced right back to the era of the Pioneer Column. During the Matabele rising against the white settlers in 1896, a mounted unit was formed in the town of Bulawayo around a nucleus of 23 expert horsemen who went on, in the words of the legendary Frederick Selous, to do 'splendid service in the supervision of the rebellion, under the name of Grey's Scouts'. Seventy-eight years later, the idea was rekindled, with a Mounted Infantry Unit emerging from an Animal Pack Transport Evaluation Team in the Inyanga mountain region of eastern Rhodesia. Led and staffed by a group of enthusiastic volunteers, the pack transport concept was soon enlarged upon along mounted infantry lines.

After trials and demonstrations had convinced the sceptics of the practical advantages of using mounted trackers in Rhodesia's bush war, the unit was deployed in September 1975.

After several successes, the unit was accorded, in July 1976, the honour of calling itself the 'Grey's Scouts' after its famous predecessors. Much was made of the achievements of this unique force after its reformation; but early successes were eventually overshadowed when the nature of the fighting revealed that the guerrilla forces could be more efficiently dealt with by the aggressive Fire Force deployments of airborne commandos of the Rhodesian Light Infantry. In 1980 the Grey's Scouts were disbanded following the creation of the state of Zimbabwe.

Above: The Grey's insignia.

GALLOPING GREY'S

Tracking the enemy on horseback during Rhodesia's anti-terrorist war, the Grey's Scouts seldom lost their quarry

DURING THE RHODESIAN War of the 1970s, a glossy army recruiting brochure described the Grey's Scouts as:

'Possibly unique in warfare today . . . a mounted infantry unit adding a new dimension of mobility in bush combat across inhospitable terrain. Every man who joins is expected to be a capable rider with a basic knowledge of equitation. The unit is self-supporting, with its own veterinary surgeons and farriers.'

As infantrymen, the Grey's were capable of conducting a variety of related tasks, operating reconnaissance, raiding, flank-protection or surveillance patrols. The unit also became valuable in psyops interface operations with the local population. But above all, the Grey's excelled at follow-up operations when used in the counter-insurgency (COIN) role.

The nature of the war in Rhodesia dictated that the Grey's Scouts had to adapt to spending long periods in the bush in their efforts to arrest the increasin[g] terrorist problem. Patrols in the vast areas of bus[h] would often continue for up to 10 days at a time, to b[e] followed by a short period of rest at a base-cam[p] before embarking on the next patrol.

With other units conducting similar COIN oper[a]tions on foot, experienced soldiers must have bee[n] tempted to volunteer for an apparently 'cushy' time [in] the saddle. But anyone who has sat astride a horse fo[r] any period of time will appreciate the hardiness [of] those who comprised the Grey's Scouts. Personn[el] were volunteers drawn from national serviceme[n,] regular and territorial army soldiers. Experience[d] riders were preferred, but in practice the gener[al] idea was that if one worked on the basis of starting o[ff] with a good soldier, one could usually sort out t[he] equitation problems later. A volunteer considere[d] suitable for the job could hopefully be taught to rid[e] properly and competently given sufficient lessons[.]

Above: Sunset patrol in the heart of bush territory. Right, top and centre: Scout and horse form the perfect tracking unit, following in the proud tradition of the original Grey's scouts (right).

was a theory that worked exceptionally well.

Training for recruits to the Grey's Scouts lasted about 18 weeks, and was carried out at Inkomo, the unit's 7400-acre base just off the main Salisbury-Sinoia road. When he joined the 'Galloping Grey's', a recruit had to satisfy his instructors that he had at least some riding potential by passing a combined training and selection course. During, and after this training, the trooper and his horse remained together so that the affinity built up between rider and mount could be used to full advantage. The Grey's Scouts were a multi-racial unit, one of many within the Security Forces. By late 1979 the Grey's were some 1000 men strong. They were an especially proud body, as is usual with any unorthodox volunteer fighting force.

The insignia of the Grey's Scouts featured the curved horn of the light infantry set beneath the proud head of a bridled war-horse. It was appropriate for the unit to honour their faithful chargers in this way. In 1978 a former commander of the unit, Major Tony Stephens, explained what the Grey's Scouts looked for in their horses:

'We stay away from thoroughbreds. They lose condition too easily and are not nearly as hardy as the crossbreeds. Ideally, we look for warhorses standing a little over fifteen hands, broadchested and strong of leg. They must not be babies, either. Five years and over make for well-matured, sensible horses who take to the sounds of war more readily and are able to complete a 10-day op without re-supply.'

But the task of tracking down terrorists meant covering many miles of rough country and a troop of Grey's Scouts might have had to travel more than 50 miles in a day. Each man had to ensure that he had an ample supply of water and rations. The latter took the form of standard 24-hour ration packs, but a wise soldier would break down his issue in order to discard most of the bulkier, heavier tins, retaining the lighter packets of dehydrated food. To replace the tinned rations, many troopers carried a liberal supply of sun-dried twists of meat called *biltong*. These tasty, protein-rich morsels weighed very little and were ideal for eating while on the move. Less weight in

rations allowed for one or two luxuries, such as an extra blanket and spare clothing. But more importantly, it meant an increase in ammunition. A rifleman can go through half a dozen magazines in a very short space of time to find himself in an embarrassing and potentially lethal situation. The few ounces made available by each tin of corned beef ditched prior to a patrol could be filled with a few more rounds of 7.62mm ammunition.

To maintain the peak condition necessary for transporting their burdensome loads, each mount had to receive the best possible care and attention. For the horse, logistics in the field tended to be conveniently simple and straightforward, with the countryside offering ample grazing and water. In the rare instances when fodder was not readily available, a bagful of locally developed, highly nutritional 'horse pellets' would suffice for the duration of a normal patrol. Although an average horse required some 80 pints of water per day, it was sufficient for an animal to drink just twice a day. The obvious advantage of this was that the Grey's Scouts were not obliged to remain close to water, which would otherwise have left them vulnerable to ambush.

The Grey's were as self-contained out of the bush as on actual operations. Farriers, men who look after the shoeing of horses, have almost disappeared from modern armies, but the Grey's Scouts had their own farriers who played a vital role in the unit, ensuring that patrols were able to switch from soft going to hard, rough country at a moment's notice without the horses going lame. A number of designs were tested before the Rhodesians came up with the ideal shoe, and thereafter, each horse was re-shod on an average of once a month.

In a contemporary interview, Major Stephens explained another aspect essential to their performance:

'With the aid of modern veterinary science it has been possible to keep disease to a minimum, and in certain cases operate in territory that horses could not have survived in in the past. Willing

Below: Training for action. A 'Galloping Grey' practises firing from the saddle at full stretch. During operations, however, the Grey's worked as mounted infantry and would rarely fight from such a precarious platform. Far right: Portrait of a Grey's Scout.

Bottom: His FN rifle at the high port, a Grey's Scout receives orders before setting out on a patrol. Mounted on horses (bottom, centre and right), the Grey's were able to patrol extensive areas and follow up any sign of enemy activity or movement more efficiently than soldiers on foot.

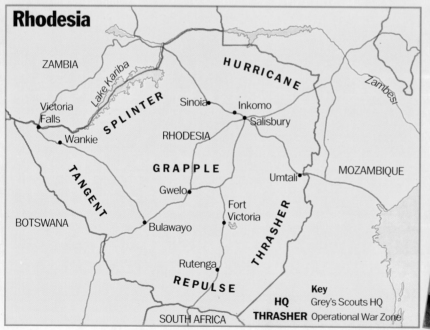

Map of Rhodesia showing:
Rhodesia

ZAMBIA, Lake Kariba, Victoria Falls, Wankie, SPLINTER, TANGENT, BOTSWANA, GRAPPLE, Gwelo, Bulawayo, Rutenga, REPULSE, SOUTH AFRICA, Sinoia, Inkomo, Salisbury, RHODESIA, HURRICANE, Zambesi, Umtali, Fort Victoria, THRASHER, MOZAMBIQUE

Key
HQ — Grey's Scouts HQ
THRASHER — Operational War Zone

veterinary officers, mainly territorial army officers and members of the unit, have launched into a massive research programme designed to improve the health and use of the horse in the field. This research includes blood chemistry which provides for a blood cell monitoring system prior to a horse being deployed on operations. A blood sample is taken from each horse and the haemoglobin level of the blood determines that the animal is not only visibly fit but is also clinically sound and not suffering from any internal parasites such as worms, which could detract from its operational efficiency. Since this programme was started the "breakdown" rate on operations has been negligible.'

Much of the saddlery was produced by unit craftsmen, cutting down on an otherwise costly outside service, and the unit also had its own transport section. Specially adapted trailers – HCVs (Horse Carrying Vehicles) – were used to transport mounts to deployment areas. Fully equipped workshops, an indoor riding school and vast areas for paddocks completed the Grey's base facilities.

By the mid-1970s, Rhodesia had been divided into six operational areas codenamed 'Splinter', 'Tangent', 'Hurricane', 'Thrasher', 'Repulse' and 'Grapple'. After being transported by vehicle convoy to a temporary base-camp, the Grey's were sent out to cover their designated

the feelings of the unfortunate Scout. In desperation he searched for his troop, riding deeper and deeper into the bush, where he soon became hopelessly lost. Despite being cut off in a potentially hazardous situation, the trooper refused to succumb to the fear he obviously felt and, determined to relocate the rest of his troop, he continued on his lonely way, picking up their spoor wherever he could. After three days

Right: Hard on the heels of the enemy. A pair of Grey's moves off at full gallop. Below: A patrol tracks enemy spoor through the bush. One of the main advantages of tracking on horseback was that the Grey's could carry up to 150lb of supplies and ammunition and were self-sufficient in the field.

section of a particular war-zone. Some troops, notably those of the fiercely aggressive Fire Force units of the Rhodesian Light Infantry (RLI), found themselves constantly in the thick of the fighting right up until the end of the war. But because the Grey's Scouts were not assault troops, actual contact with the enemy did not occur as frequently as the men might have liked. Both the RLI and the Grey's had to rely on intelligence reports regarding terrorist activity in a particular area, but while the RLI was immediately rushed to the spot, flying in on helicopters or parachuting from Dakota aircraft, the Grey's Scouts had to rely on their own ability to trail terrorist spoor through the bush, following it to contact, or until it disappeared; an often tedious and frustrating job.

Such work, however, was not without its dangers. A former member of the Grey's related to the author his experience on one such follow-up operation. His troop had been deployed after a group of terrorists in a particularly inhospitable part of the country, where the bush was so dense that it was not long before the trooper found himself separated from the main body of men. Knowing that the terrorists were not far away, it is not too difficult to imagine

he was found by his men, none the worse for his ordeal, apart from a slight case of shaken nerves. Had he known how serious a predicament he had actually been in at the time, it is doubtful that he would have maintained his relatively cool attitude in the face of the situation. It transpired that the Scouts had eventually found him by picking up on the tracks of their original quarry, the terrorists. They in turn were quite obviously tracking the solitary horseman, who was quite oblivious to the fact that he was merrily leading everyone in ever-increasing circles. That he was on horseback undoubtedly saved his life. The additional speed provided by a horse was just one of the more obvious benefits of serving in a mounted unit. At the height of the war, the Scouts' second in command, Major Williams, reported: 'You have to ride on patrol with the Grey's to fully appreciate the tremendous mobility the horses furnish ...there is no way any "terr" is going to outrun an element from Grey's Scouts!'

Another obvious benefit in conducting mounted

operations, as opposed to more conventional foot-patrols, lay in the weight factor. The foot soldier often had to cover long distances, burdened with a heavy and uncomfortable pack, while contriving to remain fresh and able to fight throughout the duration of the patrol. A Grey's Scout on the other hand, did not carry his own kit. This was distributed about his mount. The trooper only had to wear his belt-order with enough ammunition for the initial phase of a contact. His rifle, of course, was always ready and near to hand.

An additional important asset was the extra height afforded to the mounted man. A foot soldier searching for a terrorist in thick brush had to plod along with visibility reduced to a few feet and await the first shots that denoted a contact. If he were not the initial target, he could then take cover and return fire – provided he knew where the enemy was! Alternatively, he could walk at an awkward crouch, peering through the thinner bush closer to the ground in the hope of catching a glimpse of the enemy. If he was tall enough, a soldier might also be able to scan across the top of the bush, but in doing so he presented a convenient target for any watching guerrilla. But a trooper seated atop his mount had an added height advantage of about four feet, and his range of vision was increased tenfold. True, a mounted figure made for a rather large target, but movement could be so rapid and course so erratic that very few horses were ever killed.

When all of these factors were taken into account, a fighting force of considerable potential emerged. Attesting to the effectiveness of the unit is

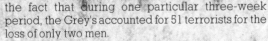

the fact that during one particular three-week period, the Grey's accounted for 51 terrorists for the loss of only two men.

In 1977 the Grey's Scouts began to operate as a squadron, as opposed to working as separate troops. This ensured a more effective combat usefulness, combined with a central element of command. The Galloping Grey's continued to function in this way until 1980, the year that witnessed the birth of Zimbabwe and signalled the disbandment of some of Rhodesia's more professional military units.

THE AUTHOR Frank Terrell served in the Royal Marine Commandos during the 1970s. A year after the completion of his service, he travelled to southern Africa where he signed on with the Rhodesian Light Infantry. The author would like to thank *Armed Forces* magazine of South Africa for permission to use material for this article.

Formed as a bulwark against the communist threat in central Europe, the mountain troops of West Germany's 23rd Gebirgsjäger Brigade are trained and equipped to fight a sustained guerrilla-type war in the upper Bavarian Alps

After skiing down a Bavarian slope in tight formation (above), jägers put their sticks to good use during a firing exercise using live rounds (top). The jägers are equally at home on the steep cliff-faces of their native land, as this soldier (far left) demonstrates. Below left: Even ice poses no problem for the highly trained men of the 23rd Gebirgsjäger Brigade – this army mountain leader is able to make a controlled descent of a glacier without the aid of ropes.

HIGH UP IN THE MOUNTAINS, and on the rocky peaks that overlook the valley floors and small villages of West Germany's Bavarian Alps, there operates a special breed of soldier – the Gebirgsjäger. 'Jäger' means hunter, a word well suited to the mountain troops of the 23rd Gebirgsjäger Brigade, a unique formation dedicated to the defence of this mountain region. The brigade is the only formation in the West German Army designed purely for mountain warfare and, as part of the 1st Mountain Division, it is capable of fighting a guerrilla-type war. The Gebirgsjäger Brigade's three battalions operate throughout the year, and comprise some of the most highly trained troops in the West German armed forces.

As a result of West Germany's geographical position, forward defence is seen as a vital component of her military strategy. Maximum combat power would have to be exerted during the initial stages of any operation, with the latter designed to minimise the damage sustained by the defending troops. In 1981, based on the latest evaluation of the communist threat in central Europe, the 1st Mountain Division was re-organised to enable it to exploit the introduction of new weapons systems. In addition, the division's component formations were reduced in size to create a larger number of more manageable units, giving the division tighter command and control.

The 1st Mountain Division's primary role is the defence of Bavaria and the Lower Alps. To this end, every soldier within the division is trained to fight in the inhospitable terrain of the mountains. As a mark of respect for their chosen battlefield, they wear the famous 'Edelweiss' emblem, a wild flower that is found high in the Alps. This badge has been associated with the Gebirgstruppen (mountain troops) since 1915. Although the division may be called upon to fight in a variety of other areas, and has the capability to do so, it is essentially a mountain formation. As such, it is endowed with the special esprit de corps that is associated with mountain troops throughout the world.

The 1st Mountain Division consists of two Panzer-grenadierbrigaden (armoured infantry brigades), one Panzerbrigade (an armoured brigade equipped with Leopard II main battle tanks), one Gebirgs-jägerbrigade (a mountain infantry brigade), a home-defence brigade (consisting of two mechanised infantry battalions and two tank battalions), an artillery regiment, an air-defence battalion, an engineer regiment and a signals battalion. In addition to these formations, there are various divisional support elements, including an armoured reconnaissance regiment and an armoured battalion.

It is, however, the jägers of the 23rd Gebirgsjäger Brigade that hold the place of honour among the Bavarians as the area's true mountain fighters. Together with the Heeresbergführen (army mountain leaders), the Gebirgsjägers specialise in high-altitude alpine combat and are specially trained and equipped to fight in difficult conditions over some of the most arduous terrain in Europe. Next to the Fallschirmjäger, the German paratroops, they constitute the only Bundeswehr army formation whose tradition dates back to before World War II. One of the reasons for this long ancestry is the Gebirgsjägers' renowned fighting ability which, coupled with its territorial recruitment, has allowed the formation to maintain strong links with its historical forebears. The brigade is based at Bad Reichenhall, five kilometres from the West German border with Austria, and is composed of three mountain infantry

MASTERS OF THE HEIGHTS

battalions: 231, co-located with the brigade headquarters; 232, based at Bischofswiesen; and 234, based at Mittenwald, near Garmisch-Partenkirchen, the home of the 1st Mountain Division.

Each of the Gebirgsjäger battalions consists of a headquarters and service support company (including an alpine platoon), three mountain infantry rifle companies (consisting of four platoons, each divided into three squads), and one heavy mountain infantry company. The latter provides the battalion's integral support, and includes a mortar platoon with six 120mm mortars, an anti-tank platoon with 21 Milan missile systems, and a field-gun platoon with six 20mm 'feldkannonen' that can also be used in the air-defence role. In addition, each battalion has a total of 76 light anti-tank weapons (Panzerfaust 3s), 478 G3 sniper's rifles (one third of which are fitted with night-vision scopes), and six four-wheel drive trucks equipped with mechanical diggers. These vehicles are a valuable asset – not only can they be used to prepare defensive positions, but they can also aid in the construction and repair of mountain tracks. This allows the Gebirgsjäger to take their vehicles, stores and equipment into the upper reaches of the mountains without having to resort to manhandling supplies until absolutely necessary. Each battalion benefits from the experience of at least 20 army mountain guides, most of whom are deployed in the alpine platoons. Their special training takes eight months, spread over the summer and winter seasons.

One asset unique to the Gebirgsjäger is the pack animal. The use of the latter to transport support equipment over difficult terrain was first exploited by German mountain troops during World War II, a period when eight mountain divisions were in existence. The Gebirgsjäger soon became renowned for the speed with which they could deploy, fully equipped for combat, from one isolated position to another, no matter how difficult the terrain. It was not surprising, therefore, that when the West German armed forces were reformed in the 1950s, the Gebirgsjäger specifically asked for pack animals to further their mobility. Despite the increasing deployment of the helicopter (Gebirgsjäger battalions are fully airportable), the mountain troops have retained a pack animal company with a complement of 54 animals. This company is commanded by an Oberstleutnant (lieutenant-colonel) who regards himself as a Gebirgsjäger first, and a veterinary surgeon second. He is responsible to the brigade commander for supplying each of the three outlying mountain infantry battalions with their quota of mules and horses.

The two types of animal employed by the Gebirgsjäger have specific roles to play. The hardier of the two animals are the mules, and each is able to carry a 130kg load. The second animal, known as the Haflin-

In December 1986, the jägers were issued with camouflage uniforms (below right) designed to replace the standard field-grey. This new form of battledress is still under trial, and is reminiscent of that worn by the jägers' counterparts in the Wehrmacht during the later stages of World War II. Below left: Using a serrated entrenching tool that has been specifically designed for cutting through compacted snow, a jäger digs a 'snow hole'. Several of these would have to be constructed if an alpine platoon was intending to effect an ambush against enemy troops. Below: Carrying up to 60 fully equipped jägers during a recent exercise in the Bavarian Alps, a Sikorsky CH-53 multi-purpose helicopter soars into the air en route to its objective.

ger, is a special breed of Austrian mountain horse. Although these can only carry a load of 100kg, they are capable of working more efficiently at higher altitudes than the mules. Each section has its own transport (mobile stalls built on Mercedes 4×4 cross-country vehicles), and each animal has its own handler. The average speed of a fully-laden pack animal column is four kilometres an hour. Bearing in mind that this includes either a 400m ascent or 300m descent, that is quite a feat.

The division's helicopters include MBB BO105s, Bell UH-1D Hueys, and Sikorsky CH-53s. The BO105 is a light helicopter capable of carrying four passengers, and, although it is used primarily for reconnaissance, it can also be used to deploy small patrols high, in the mountains. The UH-1D Huey is the work-horse of the Bundeswehr and can transport eight fully-equipped jägers. The CH-53 is the largest helicopter flown by the army and, with a capacity for 60 troops, is an ideal vehicle for larger-scale deployment. The helicopters do, however, have their disadvantages, the main one of these being their inability to fly in adverse weather conditions; a problem encountered all too often in the Gebirgsjäger's operational area. As a result of this, the mules and Haflingers have a definite advantage over the helicopter. A senior officer of the 23rd Gebirgsjäger Brigade staff had his own way of succinctly expressing it: pack animals are not grounded because of weather – and besides, you cannot eat a helicopter if

THE GEBIRGSJÄGER

Although the Edelweiss emblem has been associated with the Gebirgsjäger since 1915, the men did not establish their reputation as mountain fighters until World War II. The jägers fought in several campaigns, seeing action in the Balkans, Norway and Crete. During the Polish campaign, the 1st, 2nd and 3rd Mountain Divisions were deployed in the south. Moving in three parallel columns, they accomplished a series of outflanking moves that helped to break the Polish defences. During the campaign in Norway, the 2nd Division landed at Oslo in 1940, and marched to Trondheim where it linked up with the 3rd Division. These two units then fought their way to Narvik, seeing combat at Trofors, Eisford and Rognan. In April 1941, the newly formed 5th and 6th Divisions were sent to the Balkans to attack the Metaxas Line and open the route into Greece. The 5th Division played a major part in the capture of Crete. 1942 saw the 1st and 4th Divisions heavily committed to the drive on the Caucasus. Following the German disaster at Stalingrad during the winter of 1942, these two units were trapped on the Kuban peninsula, near the Crimea. The 3rd Division became involved in the long retreat from the Soviet Union following the Battle of Kursk in July 1943, while the 1st and 4th Divisions took up positions near Odessa. The 1st Division was later sent to northern Greece and assigned anti-partisan duties on the Albanian border.
Above: The Edelweiss insignia worn by members of the Gebirgsjäger.

you get short of food.'

The 23rd Gebirgsjägerbrigade has a total peace-time strength of around 3800 men, increased by the addition of reserves to 5100 men in the event of mobilisation. Of these 1300 reinforcements, over 75 per cent would be drafted to garrisons within a 150km radius of their homes. This ensures that the majority of reservists are familiar with the terrain of their own wartime deployment areas. West Germany has a conscript army composed of national servicemen serving 18 months, and a professional and volunteer element serving between two and 15 years. Conscripts comprise some 60 per cent of the Gebirgsjäger's total force, the remainder being professionals who man the training cadres and NCO/officer elements of the fighting units. Although the majority of its conscripts come from within Bavaria, the reputation of the Gebirgsjäger is such that a number of national servicemen from other areas of the Federal Republic volunteer to serve their 18 months with this outstanding formation. Up to 15 per cent of the division's ranks is made up of men from outside Bavaria.

The men of the Gebirgsjäger have to be familiar with the exceptional features of the mountains

The brigade has one of the broadest training spectrums within the Bundeswehr, which goes a long way towards explaining the high proportion of national servicemen who volunteer for this unit. Due to the role of the 1st Mountain Division in defending the eastern border area, and the peculiarities of the brigade's operating environment, the men of the Gebirgsjäger have to be familiar with the exceptional features of the mountains and be capable of operating in all weather conditions. To enable the battalions to train for their specialised role, they are located in areas that offer the greatest number of days per year with snow on the ground. Over 45 per cent of the training is done in mountainous terrain, and it includes the techniques of mountaineering.

Needless to say, the training is very tough. Much of the time is spent instilling and maintaining an exceptionally high standard of physical fitness, a factor vital to the success of any mission conducted in mountainous terrain. This emphasis on physical fitness is reflected in the brigade's athletic accomplishments. The Gebirgsjäger take part in various summer and winter sports at battalion, brigade and divisional level, and also compete with mountain troops of other nations.

The relationship between the Gebirgsjäger and the local community is considered important by both parties. Most of the men are local to Bavaria, an area that remained an independent kingdom until 1917, and the region is renowned for the upholding of many old traditions. One of these is the drinking of large quantities of strong local beer. In fact, the men of the 1st Mountain Division are the only troops within the Bundeswehr allowed to drink alcohol at midday. The reason behind this, according to the commander of one of the Fallschirmjäger brigades, is that Bavarians are weaned on beer!

One of the many local services provided by the Gebirgsjäger battalions is that of mountain rescue

Top left: Jägers enjoy a beer back at base camp following a patrol through mountainous terrain (top right). Far right: A jäger listens to orders while his comrades assemble (right).

ALL-WEATHER EQUIPMENT

Mountain fighting, particularly in winter, makes an enormous demand on both man and equipment, and the Gebirgsjägers are undoubtedly the best-equipped winter warfare troops within NATO. There are three types of mountain boot: medium, heavy and plastic. The heavy and medium boots are both made of leather, and, being dual-purpose, they can be used for either mountaineering or skiing. The plastic boots have been designed primarily for winter warfare and are similar to the downhill ski boots favoured by civilian skiers.

The skis themselves, adapted from the Austrian 'Apollo' models, are the best available. The bindings have three settings. The first of these secures the toe and leaves the heel free, allowing maximum mobility during cross-country patrols. The second setting uses a special clip that raises the heel and makes the gradient easier during climbing. Finally, the heel of the boot can be clamped down in the conventional manner. Coupled with 'skins', special under-ski coverings that allow the skier to move uphill without slipping back, the skis provide the Gebirgsjäger with unrivalled mobility in the snow-covered mountain terrain.

During close-quarters combat the Gebirgsjäger would switch to snow shoes, allowing far greater manoeuvrability during an ambush or attack scenario. Other winter warfare aids include Gortex under-jackets and over-trousers. These have been issued to all mountain troopers – to reduce the risk of hypothermia and increase the unit's operational efficiency in adverse weather conditions.

operations. Since all jägers are trained mountaineers and skiers, they are often called upon by the local community to venture out into the inhospitable mountains and save the lives of those involved in climbing accidents or avalanches. Apart from climbing training in summer and ski training in winter, each battalion is allocated a minimum of 100 hours of airmobile training each year. In addition, the jägers are provided with extensive instruction in the techniques of urban warfare and the storming of buildings held by the 'enemy'.

The Heeresbergführen (army mountain leaders) of the Gebirgsjäger are among the most highly trained men in the Bundeswehr. They are Unteroffiziers or junior officers below the rank of corporal who have attended and passed the long and arduous course run by the Army Mountain Warfare Centre at the Luttensee Kaserne in Mittenwald. Soldiers from other NATO forces may also attend this course, and 10 places are set aside each year for foreign nationals. Two of these are regularly taken up by members of the United Kingdom's 22nd Special Air Service (SAS) Regiment, for whom mountain warfare is a troop speciality. The course usually starts in June of each year and begins with six weeks in the town of Obereintal, in the Wetterstein area, close to the border of Austria and Switzerland. Here, in one of the Alps' highest regions, students are introduced to the ski-patrol techniques in which they are expected to become fully proficient. This is followed by three weeks at Chamonix in the French Alps, where students are taught the basics of climbing, on both rock and ice. The men then spend two weeks in the Italian Dolomites, in the area of the Cella Pass. This region is notorious for its difficult rock climbs, and

The athletic accomplishments of the 23rd Gebirgsjäger Brigade are impressive, and the unit has been consistently successful at both national and international levels. Out of 11 jägers entered at the 1984 Winter Olympics at Sarajevo, three became medallists. Below: Having adjusted the setting on their skis to provide maximum mobility during cross-country patrols, the jägers set off on yet another gruelling hike through the snow-clad Bavarian Alps.

provides instructors with an excellent opportunity to assess the progress of each soldier. One week of long, high-altitude patrols around the Watzman mountain, near to Berchtesgaden, is the next item on the gruelling agenda. By this time it is late October, and the course breaks up for leave and other military duties before reassembling for eight weeks at Mittenwald in late January/early February. Those who have not yet passed their ski-instructor's course now have the opportunity to do so. Ideal snow conditions can be virtually guaranteed during this time of year, and the course continues with two weeks in the area of Grand Paradiso in Italy. Here students learn the techniques of high-altitude ski patrolling in extreme weather. With three weeks still left to run on the course, the men spend one week around Olgoin, in southern Bavaria, where they learn 'guide' techniques. These are then put into practice during one week's training at Berchtesgaden. Finally, the soldiers spend one week in Mittenwald, where they apply the techniques of guiding to ski warfare.

At the end of the mountain warfare course, the students receive the Army Mountain Leader's Badge; a yellow and white Edelweiss upon a green oval that is surrounded by the inscription 'Heeresbergführer'. This is one of the most respected qualifications in the West German armed forces. Like all Gebirgsjägers, the army mountain leaders possess a tremendous esprit de corps and sense of professionalism that ranks as high as that of any of the soldiers within the Bundeswehr.

THE AUTHOR Peter Macdonald is a freelance defence photo-journalist and served with the British Army and Rhodesian Security Forces between 1974-80.

Known to their enemy as 'Os Terriveis' – The Terrible Ones – the men of the SADF 32 Battalion fight a tough undercover war in the bush

JUST AS THE establishment of a Marxist government in Angola had precipitated a hurried exodus of right-wing forces from that country in 1976, the creation of the state of Zimbabwe in 1980 was followed by a migration of ex-Rhodesian Army personnel across the border to South Africa. These men, veterans of 14 years of counter-insurgency warfare against Zimbabwean nationalists, were indisputably the most experienced and proficient bush fighters in Africa. It is not surprising, therefore, that the South African Defence Force (SADF) was quick to recruit many of the soldiers as mercenaries for counter-insurgency operations in Namibia (South West Africa) and neighbouring Angola. A number joined the Pathfinder Company of the SADF's 44 Parachute Brigade,

Below: 32 Battalion troopers after an airlift by SA330 Puma helicopter, and (bottom) examining target information in the Angolan bush.

THE TERRIBLE ONES

while others were attached to the SADF's specialist counter-insurgency unit, 32 Battalion.

At the time of the influx of Rhodesian troops into South Africa, 32 Battalion was based at a secret camp at Buffalo (also known as Bagani), near Rundu on Namibia's border with Angola. Few people outside the battalion were even aware of its existence at that time. The troops had been recruited largely from the National Liberation Front of Angola (FNLA), and they lived on the base with their families. Known as the 'Buffalo Soldiers', the men of 32 Battalion had been named after a group of black fighters enlisted for service at the frontier during the Indian Wars in America, and their beret badge featured a buffalo-head motif.

When 32 Battalion was formed, during the Angolan civil war of 1975-76, it was commanded by white officers and NCOs drawn from the SADF and from white mercenary recruits. Signing on for a one-year contract, an NCO could expect 11 months of service and one month's leave. Accommodation was sparse at Rundu and the NCOs were billeted in tents. For 11 months they lived and fought alongside the FNLA veterans. The battalion existed quite independently of the rest of the South African armed forces, and any contact with SADF units was discouraged. Those men of South Africa's regular army who did chance upon the base at Rundu were intrigued and provoked by the mysterious, closed air of their foreign allies, and slowly rumours began to spread about the 'uitlanders' on the Angolan border.

The shroud of secrecy lying over 32 Battalion's activities was finally lifted in 1981 when Lance-Corporal Trevor Edwards, a British mercenary who had deserted from 32 Battalion, volunteered himself for a television interview. His statements were a serious embarrassment to the South African government, for he claimed that 32 Battalion was being led over the Angolan border for counter-guerrilla operations by mercenary officers and NCOs. The SADF countered the claim by stating that every patrol leaving Rundu was led by a South African and that foreign elements were never allowed out on their own. The mercenaries of 32 Battalion, for their part, although they certainly did lead patrols into Angola, were unconcerned at South Africa's sudden ignorance of their work, seeing it just as an attempt to save face.

Today, probably as a direct result of Lance-Corporal Edwards' breach of unit security, the SADF will only accept South African citizens into its ranks. A foreigner wishing to join the armed forces will encounter considerable difficulties in doing so. Even men currently serving must now apply for South African citizenship before their contracts can be extended. It was a different matter in 1980, however, and it was not at all unknown for the battalion's former FNLA men to operate under the command of men who were previously soldiers of the Rhodesian Army.

In a typical infiltration of Angolan territory, a 32 Battalion company would be inserted either on foot or by helicopter, and a temporary base (TB) would be set up from which to mount intelligence-gathering patrols. The platoons would operate a rotating shift, enabling half of the men to patrol for two or three days while the other half rested at the TB. The base would be moved around the territory as the men scoured their area of operations. Such patrols could last anything up to six weeks, with helicopters providing supplies and evacuating casualties.

He would simply scream, 'Avance!', charging headlong into the enemy position

Following tracks over the flat Angolan bush, the patrol would eventually encounter enemy personnel. Clashes frequently occurred in the evening, when the battalion patrol and patrols mounted by the Popular Armed Forces for the Liberation of Angola (FAPLA) simultaneously converged on the same borehole to replenish their water supplies. Following contact, the units would seek to hunt down and destroy their opponents. If it was the 32 Battalion patrol that gained the opportunity to attack first, contact would be initiated by the commander of the platoon or company. The commander always led from the front, the troops invariably following the example of their leader. Usually he would simply scream, 'Avance!', charging headlong into the enemy position and firing at anything that moved. He would be hoping that his troops were right behind him. They always were.

Following Edwards' betrayal of 32 Battalion in 1981, ex-members tend to be suspicious and reticent when questioned about their combat careers. Yet, one former NCO was able to provide the following details of a routine patrol inside Angola that started in the early hours one morning.

On that occasion, the 32 Battalion patrol consisted of about 50 men commanded jointly by a Rhodesian, Corporal Walters, and a Scotsman, Corporal Hudson. (Their names are changed to conceal their identities.) The men had been infiltrated from Rundu and were 'humping 100lb bergens', their skins blackened with camouflage cream. Their uniforms, although made in South Africa, resembled those of the Portuguese in Mozambique. All items of clothing – jacket, shirt, trousers and bush hat – had been stripped of manufacturers' labels and were untraceable to any particular country. Footwear consisted of canvas and leather 'Special Ops' boots. The troops were also issued with their own style of jacket webbing, specially designed to hold the magazines of their Soviet-made AK-47 assault rifles, for they were all armed with captured Soviet or Chinese-made weapons.

In all their exterior appearances, the 32 Battalion column was indistinguishable from any Angolan unit. Should the patrol be obliged to leave any casualties behind, therefore, close inspection of their kit would

32 BATTALION

The civil war that followed the Portuguese withdrawal from Angola in 1975, ended in triumph for the left-wing Popular Movement for the Liberation of Angola (MPLA). The MPLA immediately consolidated its victory by driving its main contender for power, the right-wing National Liberation Front for Angola (FNLA), together with its white mercenary recruits, over the Angolan border into Zaire and Namibia (South West Africa). Many of the FNLA troops, thus deprived of a homeland and unwilling to subsist on charity from the West, volunteered for service with the South African Defence Force (SADF). The South Africans were glad to receive these potentially valuable, combat experienced men, and the SADF's Colonel Brëytenbach took the ex-FNLA guerrillas to Namibia where they were re-trained, re-equipped, and formed into a battalion. Originally known as Bravo Group, the unit was later redesignated 32 Battalion (its badge is shown above).
32 Battalion was organised into seven companies, each containing three or four platoons. The battalion was completed by a separate 81mm-mortar company and a reconnaissance element. Unlike other SADF units, 32 Battalion is deployed solely on counter-insurgency operations. Since its formation it has mounted highly effective operations against both the People's Liberation Army of Namibia (PLAN), which is the military wing of the South West Africa People's Organisation (SWAPO), and the Popular Armed Forces for the Liberation of Angola (FAPLA), which support PLAN in its campaign to free Namibia from South African control.

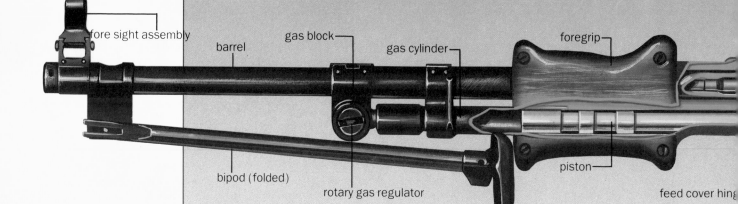

fore sight assembly

barrel

gas block

gas cylinder

foregrip

bipod (folded)

rotary gas regulator

piston

feed cover hing

THE RPD

The 7.62mm RPD (Ruchnoi Pulemyot Degtyaryev) light machine gun is one of the many Soviet-made weapons deployed by the forces of PLAN and FAPLA in Angola, and the men of 32 Battalion frequently use them for covert operations. Although the RPD has been rendered virtually obsolete in the Warsaw Pact armies by the introduction of the 7.62mm Kalashnikov (RPK), a heavier version of the AK-47 assault rifle, it remains an important weapon in the arsenals of communist Southeast Asia and of African guerrilla forces. One of a series of successful machine-gun designs by Vasily Degtyaryev, the RPD was conceived during World War II as a replacement for the DP (Degtyaryev Pakhotnyi), but manufacturing priorities allowed little progress until after the war. The gun can be distinguished by its 'club-footed' wooden butt, designed for an underhand grip with the left hand cupped under the short butt and the thumb on the inside. It also has a wooden pistol grip behind the trigger guard and a wooden hand-grip forward of the magazine.

The RPD is a gas-operated, fully automatic machine gun. It is fed with 50-round belts, each belt consisting of individual metal pockets that are linked together by short spiral springs: sustained fire is possible after linking the belts together. Alternatively, circular sheet-metal drums are used by mobile infantry, each drum containing two tightly coiled 50-round belts.

reveal no embarrassing evidence that the SADF was engaging in clandestine operations beyond South African limits. Accordingly, the SADF never released the names of 32 Battalion troopers killed in battle; they simply ceased to exist. On this patrol, a large group of 32 Battalion men narrowly escaped just that kind of oblivion.

The patrol had been fairly uneventful until the two corporals in charge acceded to their bored men's pleas and deliberately provoked an FAPLA unit. The forces grappled and then broke contact, but a resolute group of FAPLA militia then tracked the South African force over the Angolan plain and succeeded in moving up undetected on the tired troops as they lay resting in the open. The 32 Battalion column had deployed in an outward-facing U-shaped defensive configuration, consisting of three straight lines with the platoon commanders located at the two corners.

Corporal Hudson recalled that their first indication that anything was amiss came when his attention was arrested by a slight movement of some bush 'about 50m to our front'. He was still staring at this spot when a Soviet-made 40mm RPG-7 rocket suddenly exploded in their midst. The next instant the men of 32 Battalion found themselves subjected to what the corporal described as, 'the most intensive fire I have

ever experienced'. To Hudson, the very air seeme to vibrate with the concussion of countless RP(rockets, grenades and automatic fire. In his ow words, 'the air turned green' due to the volume tracer rounds coming from the FAPLA position When the two NCOs simultaneously attempted t run for the negligible cover of a solitary tree, the were beaten to it by a trooper who, the next momen was sent cartwheeling by an exploding rocket. Th 32 Battalion men were powerless to do anythin except lie absolutely flat, their bulky packs provic ing the only available form of cover.

The terrifying ambush continued for 20 minute by which time, 'everything a foot above ground leve had been sheared away, all bushes, all trees, every thing...' Corporal Hudson recalled looking back a Walters and seeing him knocked about by severa rounds. Tracer was landing, hissing and spinning only inches from Hudson's face. The perforate canteens in his pack leaked water from 'about 2

Below: In an official SADF photograph, a file of 32 Battalion soldiers moves through scorched bushland. The leading man is carrying the R4 assault rifle, the Sout African version of the Israeli Galil. Right: Unofficial photographs of white mercenaries in southern Angola, armed with the Soviet-made RPD (above) and the AK-47 (below).

Southern Africa

Lubango
ANGOLA
ZAMBIA
Rundu
Caprivi strip
ZIMBABWE
Grootfontein
BOTSWANA
Walvis Bay
Windhoek
Kalahari desert
Limpopo
NAMIBIA
Pretoria
Johannesburg
ATLANTIC
LESOTHO
SOUTH AFRICA
Cape Town

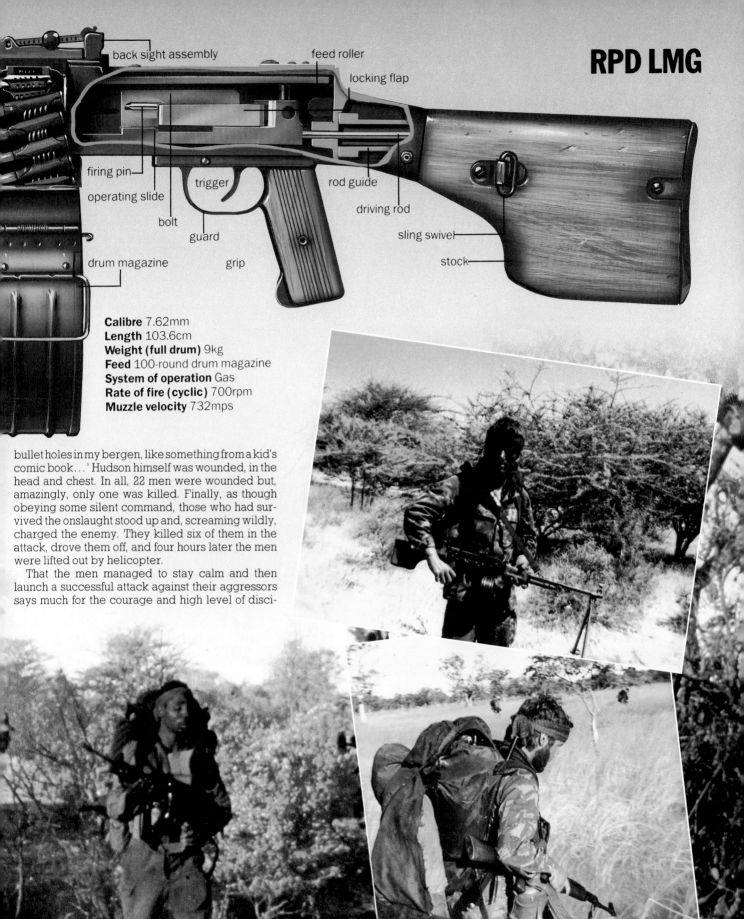

RPD LMG

back sight assembly

feed roller

locking flap

firing pin

operating slide

trigger

bolt

guard

rod guide

driving rod

sling swivel

drum magazine

grip

stock

Calibre 7.62mm
Length 103.6cm
Weight (full drum) 9kg
Feed 100-round drum magazine
System of operation Gas
Rate of fire (cyclic) 700rpm
Muzzle velocity 732mps

bullet holes in my bergen, like something from a kid's comic book...' Hudson himself was wounded, in the head and chest. In all, 22 men were wounded but, amazingly, only one was killed. Finally, as though obeying some silent command, those who had survived the onslaught stood up and, screaming wildly, charged the enemy. They killed six of them in the attack, drove them off, and four hours later the men were lifted out by helicopter.

That the men managed to stay calm and then launch a successful attack against their aggressors says much for the courage and high level of disci-

Below: A white mercenary of 32 Battalion races into position with an AK-47 assault rifle. Right: A civilian casualty, his leg wound already dressed, is carried aboard an Alouette III helicopter to be airlifted to hospital. He was injured during a firefight in which three SWAPO guerrillas were killed. Bottom left and right: In official SADF photographs, a patrol displays a stock of unearthed mines, and a Ratel-20 Infantry Fighting Vehicle is deployed against well-armed SWAPO guerrillas.

pline within the battalion. Troops in 32 Battalion learn military discipline the hard way, and even the veterans of the Rhodesian War were shocked by the ways in which the battalion officers dealt with misdemeanours of the men. For example, in one of the milder forms of punishment the unfortunate victim, wearing helmet and pack, had to manhandle a heavy truck wheel along a 36km road. If that sounds too easy, consider that the road is about a foot deep in sand and the wheel is not to be rolled but flipped over, from one side to the other.

Serious misdeeds can result in a trooper being publicly flogged over a 40-gallon drum, but in some instances even death has been meted out to maintain the respect of the men. On one occasion, a soldier was promptly shot on the parade-ground after he had aired his grievances over pay. Afterwards the CO is alleged to have asked the formed-up battalion, 'Any more complaints?' There were none.

What frustrations the men bottle up under the strict discipline of the battalion they unleash on the battlefield. Their foes in Namibia and Angola call them 'Os Terriveis', the Terrible Ones, and the communist guerrillas of PLAN and FAPLA have suffered numerous bloody defeats at their hands. For the most part, however, the details of these clashes are unknown. They occur far out in the Angolan bush, often at night, and when a report does escape the blanket of secrecy imposed by the SADF, it is usually sketchy and inconclusive.

When the SADF mounts a large, well-publicised raid over the Angolan border, such as Operation Protea in 1981, the men of 32 Battalion are usually

Top: A Soviet-supplied T-34 tank of a SWAPO fighting unit stands immobilised after an encounter with a 32 Battalion patrol. On clandestine missions in southern Angola, 32 Battalion carries such Soviet-made anti-tank weapons as the RPG-7 rocket launcher. The RPG-7, a recoilless, shoulder-fired, muzzle-loaded, reloadable weapon, fires a 2.25kg warhead that will penetrate 32cm of armour. Above: Members of 32 Battalion, wearing the brown uniform of the regular South African Army in an official photograph, uncover a cache of ammunition after a raid on a SWAPO base.

present. However, it is impossible to recognise them as they adopt the brown uniform of the regular SADF units and are thus indistinguishable from hundreds of troopers seen swarming over the Angolan plain in the photographs that are released. Very occasionally, South Africa will permit vetted journalists to report on 32 Battalion. The carefully-worded articles may then feature pictures of the unit, but the few photographs that have been released tend to depict soldiers respectably attired in regulation army uniforms and equipped with issue weapons, often strung out in a long column and captioned, 'a patrol in southern Angola'.

Cameras are forbidden amongst the troops during operations. Nonetheless, photographs exist of men in strange camouflage uniforms, bearing communist-made weapons and festooned with Chinese or Soviet webbing. The uniforms are invariably grimy, torn and tattered after weeks of wear in the inhospitable bush. The soldiers themselves are bearded and filthy and their haggard faces show the tension of constant patrolling in enemy territory. And what distinguishes some of these men from genuine Africans is that a few are obviously Caucasian, a fact that no amount of camouflage cream can disguise.

Such photographs offer strong evidence that the highly secret missions for which the unit was originally conceived are still being carried out. Recent reports from South Africa suggest that 32 Battalion was recently presented with the Presidential Colours for continuous service. This would denote that after 10 years the unit has been officially recognised, an indication that the 'Os Terriveis' can look forward to continued operational freedom in the future.

THE AUTHOR Frank Terrell served in the Royal Marine Commandos during the 1970s. A year after the completion of his service, he travelled to southern Africa where he signed on with the Rhodesian Light Infantry.

GIGN

In the early 1970s, officers of France's Gendarmerie Nationale, a militarised police force with a strength of over 60,000 men, were studying the possibility of creating a specialist anti-terrorist unit. The terrorist atrocities at the Munich Olympics in 1972 and the siege of the Saudi Arabian embassy in Paris during 1973 added impetus to their activities, and on 3 November 1973 GIGN (Groupement D'Intervention de la Gendarmerie Nationale) was formed. Originally, GIGN was divided into two commands: GIGN 1, based at Maisons-Alfort near Paris, was responsible for northern France; GIGN 2, based at Mont-de-Marsan, was assigned to watch over the south of the country. In the beginning, only 15 men, working in three five-man teams, were assigned to GIGN; overall command lay with Lieutenant Prouteau. In 1976, the northern and southern commands were merged, and the unit's strength was increased. Three years later, GIGN comprised two officers and 40 NCOs organised into three strike teams, each consisting of two five-man intervention forces, a team commander and a dog handler. Normally, one of the two officers or the senior NCO would take charge of a mission. Under normal circumstances, each of the strike forces is on full alert, ready for deployment at a moment's notice, for one week in three.

GIGN's primary role is to act as France's premier hostage-rescue unit, both at home and abroad. However, the squad is also used to deal with prison unrest, to provide protection for VIPs and to transfer dangerous prisoners between jails.
Above: GIGN's badge.

DAY OF THE SNIPER

In February 1976, a squad from GIGN, France's crack anti-terrorist force, gave an awesome display of precision shooting to secure the release of 30 children held hostage in Djibouti

SHORTLY BEFORE 0800 hours on the morning of 3 February 1976, four members of the FLCS (Front de Liberation de la Côte des Somalis – Somali Coast Liberation Front) hijacked a party of French children as they made their way from the airbase in Djibouti to their school in the town's port area. As the coach carrying the children turned onto the main coast road, the terrorists flagged it down and, once on board, ordered the driver to head south towards the village of Loyada, near the border with Somalia.

After passing through Loyada, the driver was forced to halt the coach some 180m from a Somali border post. Another terrorist then crossed the frontier in a pre-arranged move, and joined his comrades on the bus. Using the 30 children, aged between six and 12 years, and the driver as their pawns, the five hijackers then issued their demands: the immediate independence of the French territory of Afars and Issas, of which Djibouti was the capital. If their demands were not met, the children would have their throats cut.

Negotiations between the commanding officer of the 6000 French troops stationed in the colony and the hijackers began around midday. Conditions on the coach, stationed in the open under the merciless sun, quickly deteriorated, and although the terrorists allowed food and water to be served, the health of the children began to cause great concern.

As the local French negotiators sought a swift, peaceful end to the crisis, the High Commissioner of the Republic of Djibouti, Christian Dablanc, contacted senior members of the French government to discuss possible military solutions to the situation. After considering the available options, ministers decided to send a nine-man team of the Groupement d'Intervention de la Gendarmerie Nationale (GIGN), a crack anti-terrorist unit, under Lieutenant Prouteau, to the scene. Leaving Roissy airport in the greatest secrecy, the squad flew to Djibouti on board a specially converted DC-8.

On the journey from Djibouti to the Somali border, Prouteau considered the likelihood of success. His men were untried, yet they had undergone one of the toughest and most stringent selection and training programmes in the world. Then, like now, few men made the grade.

134

All candidates for GIGN are volunteers drawn from the Gendarmerie Nationale, France's para-military police force. Each man who successfully makes it through the preliminary screening process then receives a personal interview with GIGN's commander. During this critical interview, the candidate is evaluated in relation to his responses to specific questions and to an explanation of the hardships he will face while serving with the force.

Those candidates who make it through the initial stages go on to join about 100 others for the physical portion of the induction process. Endurance, agility, and marksmanship tests make up this part of the selection course. Among the physical tests are a run over eight kilometres with full combat pack, to be completed in under 40 minutes, a 50m swim in under 35 seconds, a seven-metre rope climb in under seven seconds, and various tests of rappelling ability. Candidates are also expected to be able to swim well under water. Tests of courage and skill include being pitted in hand-to-hand combat against a fully-fledged member of GIGN, all of whom are highly trained in the martial arts, or against an attack-trained dog. As excellent marksmanship is also a

Left: A GIGN marksman lines up his FR-FI sniper's rifle during an exercise. Below: Sharpshooters practise their rapid-firing drill with Manurhin handguns. Top: GIGN men pose for the camera before a SCUBA lesson. Above: Lieutenant Prouteau, the youthful and energetic commander of GIGN, in light-hearted mood.

vital anti-terrorist skill, the minimum shooting score for admission is 70 out of a possible 100 at 25m with the revolver, and 75 out of 100 at 200m with the rifle.

However, training does not end with admission to GIGN's ranks: both recent recruits and veterans are constantly and rigorously put through their paces. This diversified programme is designed to keep all members of the unit at a high state of readiness and to give them the skills and adaptability to deal with any crisis that might arise.

Because of the high pitch of physical readiness required of members of GIGN, training never stops. Both long distance steeplechase-style running and sprints are included, along with calisthenics and weight-training, to keep France's anti-terrorist team at peak efficiency. Additional training includes cross-country and downhill skiing at Barèges. Ski practice for GIGN is not only designed for physical fitness, but also to add an extra dimension to the methods by which the GIGN commandos can be inserted into an area. Along with skiing, other mountaineering skills are also taught.

As one can gather from the selection procedure, swimming also plays an important part in GIGN training. In addition to being able to swim 50m rapidly, every member of GIGN is expected to be able to swim for long distances without rest, even while pulling a 75kg dummy representing someone being rescued. Up to four hours per week are spent on underwater swimming, including both free-diving and SCUBA. One hair-raising GIGN training technique which is used to develop confidence requires unit members to dive into the Seine and lie on the bottom of the river, while huge barges pass only a few metres overhead. This exercise not only develops patience and confidence underwater, as the diver must avoid disorientation, panic and claustrophobia, but it also prepares the diver for underwater infiltration under difficult circumstances.

One of GIGN's free-diving exercises requires the swimmer to dive to the bottom of a deep ditch, read a question on a tablet, write the answer on a second tablet with a waterproof pen, and then return to the surface – all without breathing apparatus. Endurance and the ability to think quickly are developed by this exercise. GIGN men also train in 'locking in' and 'locking out' of submerged submarines as part of their SCUBA training. The various aquatic techniques are designed to prepare the commando for silent approaches to a hijacked ocean liner or other vessel. To avoid detection in such operations, GIGN has 'closed circuit' suits available which do not give off bubbles that might betray the infiltrator's presence.

All members of GIGN are parachute qualified, having attended the French jump school at Pau, and many are commando qualified as well. Since they must always be ready to carry out a parachute infiltration, each member of GIGN makes at least five training jumps per year. These exercises will normally include at least one 'wet jump' into the water, followed by a SCUBA infiltration.

A GIGN man is expected to be able to hit a moving target at ranges out to 25m

Scaling and rappelling techniques are considered of prime importance for gaining entry into terrorist-controlled buildings and, as a result, are regularly practised by GIGN. Team members are especially adept at rappelling into position, holding the rope with one hand and shooting accurately with a revolver held in the other hand.

However, all these methods of getting to the point where a terrorist incident is in progress, are only a means of putting the GIGN commander in position to do his main job: the neutralisation of his target. Since GIGN's philosophy is one of avoiding lethal force unless absolutely necessary, each member of the unit is an expert in hand-to-hand combat. Both karate and judo are studied, along with related disciplines. Quite a few members of GIGN, including Prouteau himself, have, in fact, been black belts. When GIGN men are taught karate, they normally use full contact to enhance the training benefits. Disarming and rapid neutralisation techniques are emphasised.

When more lethal force is called for, the GIGN commando has to be able to stop a terrorist or criminal immediately. GIGN firearms training is intended not only to build expertise, but also to give each unit member such confidence in handling his weapons that any 'cowboy' attitude has been eliminated. Exact shot placement is considered to be of extreme importance since any hostage-taker must be dealt with before he can harm a hostage.

With the revolver, the GIGN man is expected, as a minimum, to be able to hit a moving target at ranges out to 25m within two seconds. Additionally, each man must also be able to engage up to six targets at the same distance within five seconds. With the rifle, at a range of 200m, the GIGN sharpshooter must achieve a minimum score of 93 hits out of 100 shots. Most GIGN members can score much better, 98 with or without telescopic sights being the norm at 200m.

Below: After abseiling down the side of a building, GIGN members smash their way into a room. Right, from top to bottom: Learning the art of helicopter insertion, recruits prepare to rappel to the ground; quick-draw practice; a recruit uses a knife to hack the ice of a frozen lake.

Standard rifle drill includes firing at ranges out to 300m, though exercises may be set up that require even longer shots. Each member of GIGN averages at least two hours per day on the range, and normally fires more than 9000 rounds through his revolver and 3000 rounds through his rifle each year.

Additional range time will also be spent in practising with the sub-machine gun and fighting shotgun as well as familiarisation sessions with other weapons, including a highly sophisticated slingshot using steel balls for silent head shots! Other specialised exercises involve shooting within mock-ups of aircraft cabins and at moving vehicles. Combined with CS gas, pyrotechnics, and other such aids, these exercises train the GIGN man to score hits in the most unfavourable conditions.

GIGN members were originally armed with 9mm automatic pistols as their basic handgun, but their primary weapon is now the Manurhin 73 .357 Magnum revolver. Highly reliable, this weapon entered service in 1974. As the Manurhin remains each GIGN member's constant companion, it must be considered his basic weapon. However, each GIGN operative is also issued with his own FR-F1 sniper's rifle. The FR-F1, really a modified MAS 36, used by GIGN is in 7.62mm calibre. This bolt-action rifle has a free-floating barrel to enhance its accuracy and is fitted with butt spacers, a bipod, and a flash suppressor. Magazine capacity is 10 rounds.

The Djibouti crisis, in early 1976, was the unit's first taste of counter-terrorist work

When a sub-machine gun is needed, GIGN normally uses the Heckler and Koch MP5 in one of its many versions. The MP5A3, MP5SD, and MP5K are all used in special situations, the latter primarily in VIP protection or other covert opertions. Riot guns and various sound-suppressed weapons are also available for special operations.

As with any modern counter-terrorist unit, GIGN is also equipped with a wealth of surveillance and detection hardware. Specialist equipment includes parabolic directional microphones for listening to conversations of terrorists or other hostage-takers at a distance; thermal imagers for locating targets within a building; endoscopes (a device for obtaining 120 degree vision into a room through a tiny hole); starlight or other night-vision optics; and various other high-tech communication, detection, and surveillance devices. Pyrotechnics and explosives, including stun grenades, door openers and other frame charges, are all part of the GIGN arsenal.

GIGN's training and hardware, though impressive, are designed with one purpose: to allow GIGN to carry out its assigned missions without the loss of the hostages involved. Operationally, GIGN's record is excellent. Since its formation, the unit has rescued well over 250 hostages. These successes stem from the fact that GIGN not only functions as France's primary anti-terrorist unit, but also as a national SWAT (Special Weapons and Tactics) team called upon whenever a major crisis arises. Many crack anti-terrorist units have to wait years for employment, but GIGN sees action much more frequently. The Djibouti crisis, in early 1976, however, was the unit's first taste of counter-terrorist work. Although Prouteau's team was to carry out the actual rescue of the children, it was backed up by members of the French Foreign Legion, since there were Somali border guards close by who might try to intervene

ised that it was important to get the children out of any possible line of sight, as their presence made it difficult to align on the terrorists. Towards achieving this end, at 1400 hours on 4 February, a meal containing tranquilisers was allowed through to the children. The hope was that the children would fall asleep after eating the food, thus removing their silhouettes from the bus windows. This was precisely what happened, and at 1547 hours, all four terrorists known to be aboard the bus were visible in the snipers' sights at the same time. After patiently waiting for 10 hours for just such a moment, the GIGN snipers were given the 'shoot' order, resulting in all the terrorists being taken out simultaneously. A fifth terrorist was hit outside the bus.

A group of Somali border guards opened fire on the GIGN men almost immediately, pinning them down, but the Foreign Legionnaires gave covering fire and Prouteau with two other men rushed towards the bus to free the children. Another terrorist had boarded the bus under the covering fire from the Somalis, and he managed to kill one little girl before being cut down by the GIGN assault force. The girl was quickly avenged, however, as the Legionnaires and GIGN containment force poured withering fire into the border post and its garrison. It was reported that the leader and planner of the terrorist attack was killed during this engagement.

The Djibouti operation was a classic hostage rescue mission: the men of GIGN had to travel to another continent with little notice, quickly gather intelligence in a hostile environment, plan an attack, wait patiently for the proper moment to strike, and make every shot count when the order to fire was finally given. Though one hostage was lost, the operation was a success in that 29 other children were saved. By their prompt action, GIGN had sent out a clear signal to

GIGN first shot to prominence in February 1976, when members of the Somali Coast Liberation Front hijacked a coachload of children in Djibouti. Flown out at a moment's notice, Prouteau and nine of his men took up positions around the coach, and using FR-F1 rifles, dealt with the hijackers simultaneously. Above left: The scene on the coach after the release of the hostages. Although the children spent little more than a day on the bus, conditions on board deteriorated quickly, with many of them suffering from heat exhaustion and stomach cramps. Left: One of the lucky survivors of the crisis walks to freedom. Above right: Smiles all round as members of GIGN return to France.

against any rescue attempt. After carrying out a reconnaissance of the area, Prouteau established his command post in a palm grove close to the hijacked bus and placed his nine marksmen, all armed with the FR-F1 sniper's rifle, at advantageous sites around the target.

GIGN tactical doctrine in such a situation called for Prouteau to be in constant radio contact, using throat microphones, with the marksmen. Before moving into position, each shooter had been assigned to watch over a particular portion of the bus and, to ease the flow of information, each terrorist had also been given a recognition number. Using this system, each of the marksmen could instantly let the commander know when he had his particular terrorist in his sights just by giving the target's number. Since all the terrorists had to be eliminated at the exact same time to avoid a general massacre of the children, Prouteau decided that he would only give the 'shoot' order when all of his sharpshooters had a clear view of the hijackers.

When planning the rescue attempt, Prouteau real-

those contemplating terrorist acts against France: they could respond to any outrage with lethal finality.

In the years since Djibouti, GIGN has been used on other well-known operations such as the attempted prison break at Clairvaux prison in January 1978, in which GIGN marksmen once again saved several hostages through precision shooting. As are the SAS and Germany's GSG9, GIGN is used overseas to foster France's diplomatic interests by training and assisting foreign anti-terrorist units and VIP protection groups. Perhaps the most famous instance of their deployment in this role occurred in 1979 when members of GIGN helped train the Saudi Arabian National Guard for the operation to retake the Great Mosque which had been occupied by fanatics. However, GIGN has trained many other units, especially those in France's former colonies.

GIGN's skills have remained sharp through exchange training with other Western anti-terrorist units and through employment on high-risk assignments within France. 'Gigene's' ready team, sitting at Maisons-Alfort waiting to move into action as these words are being written, remains one of the world's most formidable counters to chaos.

THE AUTHOR Leroy Thompson served in Vietnam as a commissioned officer in the USAF Combat Security Police.

Djibouti
GIGN, February 1976

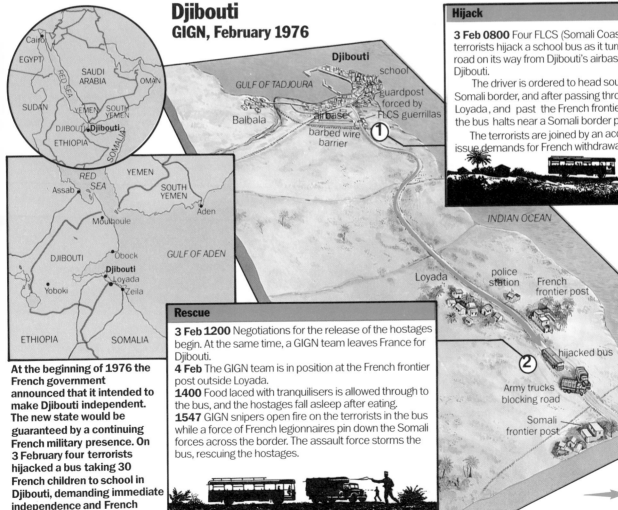

Hijack

3 Feb 0800 Four FLCS (Somali Coast Liberation Front) terrorists hijack a school bus as it turns onto the coast road on its way from Djibouti's airbase to a school in Djibouti.

The driver is ordered to head south towards the Somali border, and after passing through the village of Loyada, and past the French frontier post beyond, the bus halts near a Somali border post.

The terrorists are joined by an accomplice. They issue demands for French withdrawal from Djibouti.

At the beginning of 1976 the French government announced that it intended to make Djibouti independent. The new state would be guaranteed by a continuing French military presence. On 3 February four terrorists hijacked a bus taking 30 French children to school in Djibouti, demanding immediate independence and French withdrawal.

Rescue

3 Feb 1200 Negotiations for the release of the hostages begin. At the same time, a GIGN team leaves France for Djibouti.
4 Feb The GIGN team is in position at the French frontier post outside Loyada.
1400 Food laced with tranquilisers is allowed through to the bus, and the hostages fall asleep after eating.
1547 GIGN snipers open fire on the terrorists in the bus while a force of French legionnaires pin down the Somali forces across the border. The assault force storms the bus, rescuing the hostages.

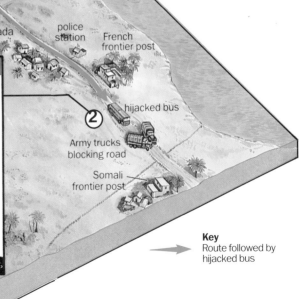

Key
Route followed by
hijacked bus

The British South Africa Police (BSAP) of Rhodesia – not to be confused with the police force of South Africa – was raised in 1896 to combat insurgent elements among the Shona and Ndebele tribes. From these 19th Century military beginnings, the BSAP developed into a civil police force and became Rhodesia's 'senior service', over the army and air force. During the early stages of the war the BSAP was responsible for the majority of counter-insurgency operations and was supported by the army and air force. Guerrillas were regarded as politically motivated criminals and, when captured, were brought to trial and convicted through normal judicial channels. Counter-insurgency operations, however, demanded very different skills and tactics to normal police work and PATU was established in 1966 to carry out these duties.

As the war escalated, the combat role of the BSAP was expanded and several new units were formed for special duties in the anti-terrorist role. These included the highly successful Support Unit, known as the 'Black Boots', the SB-Scouts, a Special Branch unit of agents for intelligence gathering, a Police Mounted Unit to increase police mobility, and, as the war spread from the countryside into built-up areas, the Urban Emergency Unit tasked with combating urban terrorism.

Above: The PATU badge, the design of which was based on a lion's spore, was awarded to PATU men after they had proved themselves on active operations.

PATU, the Rhodesian police anti-terrorist unit, waged a tough campaign against infiltrating guerrillas in the Zambezi valley

ON 17 MAY 1966 a white farming couple, Mr and Mrs Viljoen, were killed on their Nevada Farm, situated near Hartley in central Rhodesia. The Viljoens were the victims of a raid that followed a large incursion by ZANU (Zimbabwe African National Union) guerrillas across the Zambezi river from neighbouring Zambia in late April. Soon after the guerrillas had crossed the border, part of their force had been located by the security forces on the outskirts of Sinoia and elimin-

reservist force, but one man in particular, Bill Bailey, a regular officer with the BSAP, was far-sighted enough to realise that the threat posed by the unconventional style of rural warfare favoured by the guerrillas, would have to be met with a more specialised response than the ordinary pattern of everyday police work.

Bailey himself had a great deal of experience of special-operations work: during World War II he had served in the Western Desert with the Long Range Desert Group (LRDG) and had then gone on to work with partisan groups in Albania.

In 1964, realising that the growing terrorist threat might take hold of his own district of Lomagundi, which lay between the Rhodesian capital, Salisbury, and the Zambezi crossing points, Bailey established

WAR IN THE BUSH

ated by a combined police, army and air force operation, code-named Pandora; but the gang responsible for the killing of the Viljoens had got through. The Sinoia and Hartley incidents were to prove a turning point in what was now beginning to look like a war in Rhodesia.

In the early 1960s trouble had already begun to ferment in Rhodesia's African townships and by 1963 this unrest was augmented by terrorist attacks mounted by dissident nationalist guerrillas who were based and being trained outside the country. At first, these raids were on a modest scale and not particularly successful, but by 1966 the guerrillas were better organised and equipped, and had begun to launch large-scale incursions into the country. To combat the growing security problem, the British South Africa Police (BSAP) were recruiting a police

a small military-style outfit within the police known as the Tracker Combat Teams. Senior police officers in Salisbury, however, took a very dim view of this paramilitary approach and Bailey was ordered to disband what they regarded as his 'private army'.

But Bailey kept his concept alive and his team members volunteered to a man for the officially-tolerated Volunteers for Advanced Training (VATs). Known in those days as 'Bailey's Bloody Bush Babies', the VATs, most of them police reservists, concentrated on really getting to know their areas. Many of them were farmers and, working in pairs, they checked out their own farms and the surrounding territory, keeping a sharp eye open for evidence of terrorist movement in the area. Knowledge of the terrain, the local villages and the people, Bailey knew, was the key to counter-insurgency work.

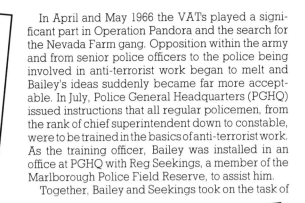

In April and May 1966 the VATs played a significant part in Operation Pandora and the search for the Nevada Farm gang. Opposition within the army and from senior police officers to the police being involved in anti-terrorist work began to melt and Bailey's ideas suddenly became far more acceptable. In July, Police General Headquarters (PGHQ) issued instructions that all regular policemen, from the rank of chief superintendent down to constable, were to be trained in the basics of anti-terrorist work. As the training officer, Bailey was installed in an office at PGHQ with Reg Seekings, a member of the Marlborough Police Field Reserve, to assist him.

Together, Bailey and Seekings took on the task of

PATU volunteers train with a light machine gun (top) and FN FALs (above). Weapon instruction formed part of an SAS-style regime of back-breaking training instituted by Inspector Reg Seekings. Above right: PATU stalwarts take a break during a patrol. On the left is Bob Mansill and in the centre, Herby Gibbon, winner of the Mr Rhodesia and Mr South Africa body-building awards. Right: Jerry Cleveland of PATU prepares rations in the bush.

forming a new police section, specifically trained and organised to meet the terrorist threat, and on 1 August 1966 the Police Anti-Terrorist Unit (PATU) officially came into being. A desperate shortage of equipment and facilities, however, meant that only 32 volunteers could be accepted at the outset. Only the men who passed a rigorous selection course devised by Bailey and Seekings would be taken on.

Seekings, who had served with the British SAS from its very beginnings to the end of World War II, had been a close friend of Bailey during the desert days and, together, they resurrected the notoriously demanding style of training that has always been the hallmark of the SAS. 'Reg's PT', as this regime soon became popularly known among the PATU volunteers, was designed to cream off only the very best from the ranks of the police reserve. Seekings' approach was quite straightforward:

'You've got to hammer these people in training. I'll always remember Jock Lewis, going back to the SAS days, saying, "Training must be made as hard as is humanly possible, so that when you go on an operation you find that it's easier than your training." It gives a man confidence and a confident man, with a little bit of luck, will always come through. It was very hard to put across all those drills, and I had to drive it in and be really set in my ways, because it was hard to make a man understand that I wasn't asking him to win a VC. It was to save his own bloody life. He must attack without question, without hesitation, go in.'

With minimal equipment, a rucksack and a rifle each, scrounged from the quartermaster, the first volunteers began their basic training. Seekings describes the qualities he was looking for:

'We took them out into the field to see how a man reacted physically, how he could find his way around, how good he was at bushcraft and how he was going to fit into the picture. How was he going to get along with his mates? Could he mix in? Could he take authority? And could he keep his mouth shut? The biggest thing in all this was making a man keep quiet. When you're in the bush you've got movement going on all the time. Elephants and other animals are moving around, but once you get the sound of metal, the chink of a rifle, or a voice, the game's up.'

Having ascertained whether a man was suitable material for PATU advanced training, Seekings then sent the selected recruits on an endurance course and gave further instruction in weapons handling and map reading. The trainees were deliberately taken out to areas where they had to cope with all the hardships of rough terrain, extreme heat, and water shortage. Survivors of advanced training were selected to make up the operational PATU teams.

Bailey and Seekings took the training regime on a tour of the police districts and within six months PATU was operational. But PATU still had more than just the enemy to contend with. Despite PGHQ's acceptance of the necessity for anti-terrorist personnel after the Sinoia and Hartley incidents, many officers continued to view PATU's existence as a radical departure from the traditional values of police work and found the emphasis on weapon proficiency particularly hard to swallow. PATU also met with considerable animosity from within the ranks of the regular army, who regarded the PATU men as 'coppers' who should stick to police work and leave the task of hunting down armed and dangerous terrorists to men who knew what they were doing. Subsequent events were to alter these opinions.

Above: PATU members receive their first issue of camouflage uniform. This picture was taken during a PATU operation and legend has it that the quartermasters were far from happy at having to enter the war zone. Right: Inching forward through thick bush, a PATU stick moves into the attack. Below: A very early photo of PATU men, taken before the issue of camouflage and when FNs were in very short supply. Several of the group are armed with old .303s. Far right: Inspector Reg Seekings, the driving force behind the PATU training regime.

While PATU weathered this friction and protest undeterred, such opposition made it hard for the teams to get hold of the weapons and kit they needed for their anti-terrorist role. At the outset, there were very few modern FN rifles to be had and many men had to make do with old bolt-action .303s, a fine weapon but not well-suited to close-quarters contacts with an enemy fielding AK automatic assault rifles, RPD and RPG machine guns. Uniform was also a problem. Camouflage kit was not forthcoming and the patrols were forced to take to the bush in police 'riot blues'. Only when it was found that many of the guerrillas operated in similar blue clothing were camouflage jackets issued.

It was on this precarious footing that the first PATU patrols went into action in the Zambezi valley, right on the front line where the guerrillas made their incursions. The patrols, organised into sticks of five men made up of four Europeans and one African, were dropped off at police stations in the area and then escorted down into the valley by local police officers. Down in the valley it was a war of nerves. Reg Seekings, who participated in many of the operations, describes the conditions PATUs faced:

'You could have a flurry of contacts and then there would be long periods with no contact at all. That was the trouble. It was really ball-aching for the chaps going in, and then there would be a new incursion with new groups coming into the country and all of a sudden the place would come alive.

'The plan was for us to be seen as often as possible on the river banks by the opposition, the

INSPECTOR REG SEEKINGS

In January 1939 Reg Seekings joined the British Territorial Army, mainly for the boxing, and that year won the East Anglian light-heavyweight championship. When war broke out, he volunteered for sea raiding operations and was attached to No. 7 Commando.

After training in Felixstowe and Scotland, he was sent with Layforce to the Middle East, where he took part in several missions, including the seaborne landing at Bardia. When Layforce was disbanded, Seekings was recruited by David Stirling as a founder member of 'L' Detachment of the SAS and went on to serve with great distinction in 1 SAS in North Africa, Italy, France and northwest Germany.

After the SAS was disbanded in late 1945, Seekings spent several years as a publican in Ely before emigrating to Rhodesia, where he managed a tobacco farm and a chain of stores.

In the early 1960s, as violence flared in Rhodesia's African townships, the government decided to strengthen its security forces by creating a police reserve. Seekings joined the Marlborough Field Police Reserve and became a section leader. When armed incursions began in 1966, he joined forces with an old Long Range Desert Group friend, Bill Bailey, who was the Chief Superintendent of the British South Africa Police, to form the Police Anti-Terrorist Unit (PATU). Seekings became the unit's chief instructor. He served with PATU until its disbandment on 31 July 1980, when he retired from police duty.

idea being to stop them having a clear run. Instead of having a clear passage across the valley at night, if they saw us here, there and everywhere, they would never know where we were going to be and they had to proceed with caution. The longer it took them to cross that valley, the less likelihood there was of them being an active force when they reached the other side. They'd be physically buggered and short of food and water, and suffering. Mentally, too, they'd be in trouble and that was when we picked them up.

'We had cleared the Zambezi valley of what local population there was and it became a killing area. Anything we saw in there, we knew it was a terrorist group. During the day we'd move down to the river and be seen on the banks. Then we'd move out again and be seen a few miles further up, back on the dirt road checking for new spore. The whole idea was that they knew that the area was being patrolled, and they couldn't tell whether it was just one or a dozen patrols.

'In a 100-mile area, we'd have only one stick operating and we even went to the extent of changing the dress of our patrols. We'd turn our hats inside out and change the make-up of our kit and the way it was slung. They couldn't distinguish our faces. At times they attacked us, but their main aim was to get through, so it was a case of us attacking them.'

For several years PATU fought this dangerous game of cat-and-mouse in the sweltering heat of the Zambezi valley, operating without army and air force support. By 1970 PATU numbered some 1000 men distributed around the country. Their actions in the

Above: Bill Bailey, the ex-LRDG Rhodesian police officer who first introduced the idea of a police anti-terrorist unit. Below: A PATU stick. Second from right is Section-Leader Campbell-Watt who was once ambushed by mistake by an army patrol, but wiped them out before the error was discovered. Bottom: PATU men with a personnel carrier in the bush.

border areas earned them a great deal more respect from the regular army, and PATU groups would often operate in conjunction with the military. Essentially, PATU was a tracker/reconnaissance force, and if a large contact was made, the stick was supposed to call in the army and helicopter support in the form of the Rhodesian Light Infantry (RLI) 'fire forces'. Frustration at this procedure, however, often got the better of the PATU men and, depending on the level of indoctrination with police procedure and the characters of the men involved, they would 'do a Nelson' and take on the guerrilla group themselves.

Sticks were deployed to protect convoys carrying vital supplies

In the early 1970s the guerrilla war escalated sharply, and by 1973 the guerrillas had firmly ensconsed themselves in the north and northeast of the country. The bush war had also begun to spread into other areas of the country and PATU operations had extended to the southeastern border areas where Rhodesia met Mozambique. As PATU zones of operation expanded, so did its role. In the south, sticks were deployed to protect convoys carrying vital supplies from South Africa to an economically blockaded Rhodesia, and all over the country PATUs' regular police training proved invaluable for the gathering of intelligence as Reg Seekings outlines:

'At this time, the enemy had infiltrated a lot of political commissars who were holding meetings in the villages. PATU was used to try and break up these meetings, but we found this very difficult. The problem was trying to separate the terrorists from the locals and there was always a big danger of innocent civilians getting hurt. It was a very thankless task. We eventually got down to trying to ambush the terrorists as they left the village after the meeting but that also had its problems. Most villages had a number of exits so selecting the ambush point was a bit hit or miss. Also, if the contact was initiated too soon, the terrorists would not hesitate to use the villagers as a fireguard.

'Another job we had to do was to find out who was feeding them and where they were feeding, so that we could set up ambush points at these places. We tried to get round to villages that had suffered atrocities at the hands of the terrorists and find out from the villagers the whereabouts of the terrorists, their habits, and also get a description of them. All this information was then passed to all the army personnel and patrols in the area.'

During the latter stages of the war in Rhodesia, PATU's role became closely integrated with that of the military. Although they were still policemen, many of the barriers that had isolated them from the armed forces had evaporated and they were widely respected for their experience in the field, their ability to beat the guerrillas at their own game and their professionalism in action.

Although the war in Rhodesia had spawned several highly professional anti-terrorist units such as the RLI fire forces and the Selous Scouts, PATU had been in the thick of the action since the very beginning and had steadfastly maintained that when faced with a threat to internal security, the police had a major role to play. Their actions proved that point.

THE AUTHOR, Jonathan Reed and the publishers would like to thank Reg Seekings, former chief training officer of PATU, for his invaluable assistance in the preparation of this article.

HOLLAND'S MARINES

One of the oldest military formations in the world, the Dutch Marine Corps was founded on 10 December 1665.

The modern Royal Netherlands Marine Corps numbers some 2800 men. Ninety per cent of the men are regulars, with officers and NCOs serving for seven and four years respectively; 10 per cent are national servicemen. At the end of their stint, the conscripts can either sign on as regulars or join the reserves.

The corps' strength is divided between two regional commands: Netherlands, allocated to NATO, and the Dutch Antilles, based on the islands of Aruba and Curaçao in the Caribbean. The NATO command consists of the 1st Amphibious Combat Group (1ACG), some 700 men based at Doorn; two 'quick reaction' forces, ready for action anywhere in the world at a moment's notice; and the 7 Netherlands Special Boat Section. Based on HMS *Thetis,* the latter unit's specific tasks are long-range reconnaissance and anti-terrorist operations on oil rigs.

One other marine unit that is based in the Netherlands is the BBE (Bizondere Bystand Eenheid – literally, the 'Different Circumstances Unit'). Based at the Van Braam Houckgeest barracks, the BBE is a full-time counter-terrorist outfit and its ranks are filled with specially trained marines.

Above: The badge of the Royal Netherlands Marine Corps with the unit's motto – Quo Patet Orbis (the Whole World Over).

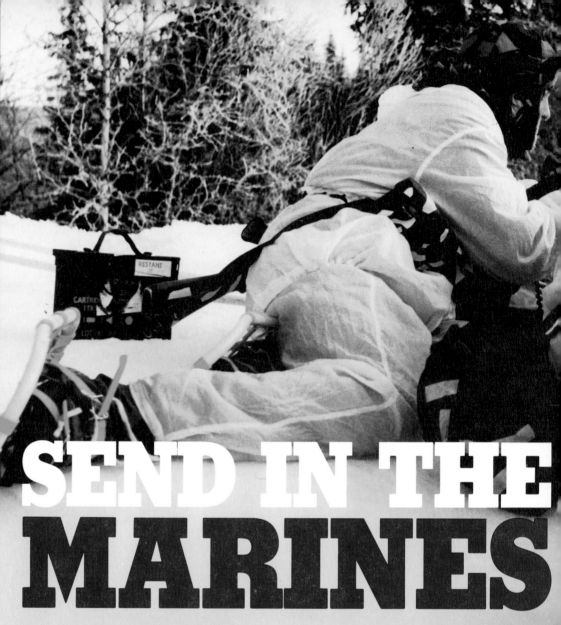

SEND IN THE MARINES

After 19 days of siege negotiation, the Dutch authorities unleashed the Marine anti-terrorist force against gunmen holding a hijacked train

ZERO HOUR minus 30 minutes. Under cover of darkness in the early hours of 11 June 1977, the covering platoon of the Royal Netherlands Marine Corps inched its way forward to the pre-set fire positions around the carriages of the hijacked train. Once in place, the men sighted their GPMGs and FN FAL rifles on the target, ready to unleash a deadly accurate curtain of fire to isolate the 13 South Moluccans from their 80 or so frightened hostages. Elsewhere around the train, sniper marksmen from the Dutch RijksPolitsie also took up position. These men, as highly trained as the marines, carried Heckler and Koch 7.92mm rifles adapted to a sniping role and equipped with a single-point sight: one that focuses a red dot on the target. Although fearfully accurate, the rifles were also a psychological weapon. If a hijacker saw the dot on his body, he would either freeze or try to take immediate evasive action – a terrorist trying to hide from a sniper has no time, no thought to start killing his captives.

The crisis faced by the Dutch authorities had started at 0830 hours on 23 May, when the South Moluccans seized control of a commuter train on its way between Rotterdam and Groningen in northern Holland. Halting the train at De Punt, the hijackers then released some of the passengers on board, chained and padlocked the carriage doors, and blacked out the windows with newspapers. Next, they issued their demands for the safe release of the hostages: the Dutch government had to exert pressure on the Indonesians to grant independence to their homeland of South Molucca. The hijackers also demanded that several of their comrades held in Dutch prisons for earlier offences should be released and that a Boeing 747 be made available at Amsterdam's international airport, Schipol. To prove that they meant business, the hijackers shot and killed the train driver, throwing his body onto the tracks.

Negotiations between the South Moluccans and the authorities were to drag on for more than two weeks, but even at this early stage, the Dutch were considering a military solution to the situation. Within hours of the take-over, 24 marines of the BBE (Bizondere Bystand Eenheid – Different Circumstances Unit) had left their barracks for De Punt, where a tight military cordon had been thrown around the train. Their first task was to gather intelligence on the South Moluccans: their number, positions and weapons. It was agreed that food trolleys could be taken out to

Left: On exercise in Scandinavia, two men of the Royal Netherlands Marine Corps give covering fire from behind their overturned sledge. One of the oldest in existence, the corps has developed close ties with the British Royal Marines and is likewise committed to the defence of NATO's vulnerable northern flank. However, the Dutch Marines also field a specialist anti-terrorist squad.

Those men who volunteer for service with the Royal Netherlands Marine Corps have already completed their basic training but, before they are accepted into the anti-terrorist unit, each man is given a stringent psychological examination to assess his suitability. In particular, his behaviour in action and his reaction to stress are closely studied.

Once over this hurdle, each recruit is put through his paces during a training course that lasts for some 48 weeks. In the first stage, lasting 16 weeks, he is attached to the unit's third squad and, if successful, will then be placed in one of the two active squads for further on-the-job training and assessment. The first part of the programme concentrates on developing specialist anti-terrorist techniques; the second stage continues this and also tries to promote a recruit's ability to work as part of a close-knit team.

Much of the marine training is similar to that of other anti-terrorist squads: close-quarters fighting, storming buildings, marksmanship and hostage rescue. However, the recruits are also taught the finer points of riot control, and maintain especially close links with other Dutch security forces.

To give their training a practical edge, recruits are taken to Schipol, Holland's main international airport, for exercises and also learn the lay-out of other major airports. Like the SAS and West Germany's GSG9, they are taught the techniques for storming hijacked aircraft. The trainees also spend a large part of the course on rifle and pistol ranges improving their marksmanship; machine-pistols and hand-guns are the squads' main weapons. However, the men are also taught to use standard NATO weapons as well as infra-red sights and image intensifiers.

After passing the training course, each man usually remains with the unit for two years.

Dutch Marine 1970s

Armed with a 9mm Uzi, this man, a member of the corps' anti-terrorist section, is wearing an armoured flak jacket over his 'woolly pully', combat trousers and a US M1 helmet. His webbing includes a haversack for grenades, a pouch for medical equipment and a US-type water-bottle cover. He also carries a fighting knife and a revolver.

the train by the Red Cross, who would have to strip naked to prove that they were unarmed. None of the hijackers seemed to notice that the Red Cross personnel were young and muscular; in fact, they were policemen trying to gather useful information.

Any authority faced with a hostage crisis has to recognise one brutal fact: if the hijackers start to kill their hostages in the early stages of a siege, there is little that they can do about it. There was no point in charging in, guns blazing, until they knew as much about the South Moluccans as possible. Gathering intelligence takes time and most sensible governments negotiate for as long as possible.

As part of this process, the authorities will always try to secure the release of some of the hostages – usually the sick, women and children. There are two reasons for this: an obvious desire to save lives without bloodshed, and a need to get up-to-the-minute intelligence. On the fourth day at De Punt, the negotiators managed to obtain the release of a pregnant woman who was able to provide a great

In the late 1970s members of Holland's South Moluccan community carried out a series of acts to highlight their demands for an independent homeland, free from Indonesian interference. In December 1975 two groups seized control of a train near Beilen (main picture) and the Indonesian consulate in Amsterdam (below far right). In both cases, the authorities refused to yield to the South Moluccans, and the hostages were eventually released. Nearly two years later, on 27 May 1977, a school at Bovensmilde was taken over (below right). Faced with a government unwilling to agree to their demands, the South Moluccans freed their captives on the 27th.

deal of intelligence about the South Moluccans: what weapons they had (automatics and grenades), who were the leaders, where they slept at night, how many were on guard at a given time and where they were stationed. Because hostages often begin to identify and even sympathise with the hijackers, a process known as the 'Stockholm Syndrome', this information had to be collected as quickly as was feasible. A hostage freed some time after the hijacking has taken place may well give deliberately false or misleading details.

In fact, talks between the authorities dragged on until 9 June. A South Moluccan doctor initially led the negotiations for the authorities, but as time went on his true loyalties became suspect – his natural sympathy for his own people might lead to his giving the hijackers information about the forces ranged against them. Matters were brought to a head on 10 June. The hijackers remained steadfast in their demands, which the Dutch could never agree to.

The decision for the BBE to go in had already been made; a decision that the Dutch Minister of Defence let slip in an interview with journalists, 'We've reached a stalemate…maybe it's time for other

measures.' Unfortunately his prophetic words were quoted on the radio and, despite the electronic jamming intended to isolate the South Moluccans from the outside world, they managed to hear one of the broadcasts that quoted the minister's words.

By the 10th, the rescue teams were ready to storm the train. The Dutch SBS had been spending a good deal of time infiltrating along the tracks below the train; always at night when there was little chance of either the hijackers or potential sympathisers being aware of their activities. During their frequent sorties, they had planted explosives on the tracks in front of the train, to be used as a noisy diversion when the attack finally took place, and had placed sensitive listening devices on the body of the train itself. The teams had also been provided with psychological profiles of the South Moluccans, courtesy of the Dutch Intelligence Service. The authorities had a pretty good idea of how the hijackers would react if they realised that their demands were not to be met: they were amateurs, and amateurs were judged likely to start shooting indiscriminately.

At first light on the 11th, the BBE marines prepared for action. They were formed into two bodies – one to

FIGHTING FOR A HOMELAND

Following the creation of the United States of Indonesia out of the former colony of the Dutch East Indies in late 1949, the new government attempted to impose its will over the islands of South Molucca. In response, the Moluccans proclaimed an independent Republic of the South Moluccas, and the confrontation escalated into armed rebellion. The rising was suppressed and in the aftermath some 15,000 Moluccans were transported to the Netherlands where, housed in refugee camps, they faced major problems of assimilation.

Resistance to the Indonesian authorities continued from Holland. In 1975, the Free South Moluccan Youth Organisation (Vrije Zuidmolukse Jongeren – VZJ) decided on a policy of direct action: Indonesian targets in Holland were attacked and there was an unsuccessful attempt to kidnap Queen Juliana.

In late 1975, however, two attacks by members of the VZJ gained international attention. On 2 December, seven Moluccans hijacked a train near Beilen, killed the driver, and demanded the release of Moluccan prisoners and talks on the future of their homeland. Two days later, another six Moluccans seized the Indonesian consulate in Amsterdam, again taking hostages. Despite the death of one of the train hostages on the 4th, both groups gradually released their prisoners and the crises ended on the 14th and 19th of the month.

On 23 May 1977, two groups carried out co-ordinated attacks on a train at De Punt and a primary school at Bovensmilde. In both cases, several dozen hostages were seized and, although the schoolchildren were released unharmed on 27 May, the Dutch ordered specialist marines to storm the train. On 11 June, six Moluccans and two hostages were killed during a

shoot-out, but the marines suffered only one minor casualty.

Another incident involving members of the VZJ occurred on 13 March 1978, when three Moluccans took over government offices in Drenthe, near Assen, and demanded the release of their comrades serving long prison terms for their activities, in exchange for the lives of their 71 hostages. After the deaths of five civilians and one hostage, the siege was ended on the next day when Dutch marines stormed the buildings.

The authorities faced the succession of crises with a pragmatic mixture of harshness and compassion: those directly involved in the outrages received stiff prison sentences, while the South Moluccan community received government aid to improve its conditions in Holland. The approach appears to have been successful. VZJ leaders have voiced the view that hopes of an independent South Molucca should be abandoned.

Key
⭐ South Moluccan terrorist incidents
— Railways

NORTH SEA

Groningen
23 May 1977 De Punt
23 May 1977 Assen
Bovensmilde
Beilen **13 Mar 1978**
2 Dec 1975
DRENTHE

Amsterdam **4 Dec 1975**
Schipol airport ⭐

The Hague
Rotterdam
Utrecht
Arnhem

NETHERLANDS

WEST GERMANY

Eindhoven

BELGIUM

RNAF Starfighters
BBE, Royal Netherlands Marines

Hijack
23 May 0830 Terrorists take over Rotterdam-Groningen train and force the driver to stop near De Punt.

Siege at De Punt
May-June 1977
After a siege lasting more than two weeks, the train at De Punt near Groningen held by South Moluccan terrorists was stormed by the crack troops of the Dutch Marine anti-terrorist squad.

Starfighters
11 June 0453 As a team of marines prepares to storm the train, a flight of F-104s screams overhead with afterburners alight.

① To Groningen

carriages occupied by hostages and terrorists

②

③ To Assen

Assault
The marine assault force storms the train using explosive charges to blow the doors open.
0500 The train is secure.

storm the train and the other to provide covering fire. Aside from their obvious function, the cover platoon was to lay down a curtain of fire to isolate the South Moluccans from their captives by a stream of bullets fired into the carriage, where listening devices had indicated that the South Moluccans spent much of their time. The attack platoon was equipped with Heckler and Koch automatics and Smith and Wesson police special pistols; they wore regular combat uniforms with body armour; their faces and hands were smeared with camouflage cream.

Two thoughts were uppermost in the men's minds: the fear that one or more of the hostages might be killed before they could take out the hijackers, and the possible reaction of the hostages to the rescue. Electronic surveillance had indicated that several captives had fallen prey to the Stockholm Syndrome and were identifying with the South Moluccans. Because of this development, the marines knew that they would have to go in very quickly, with a great deal of noise, to prevent those hostages suffering from the syndrome hindering the rescue attempt.

Moments before the 'go' order, the marines finished their pre-assault preparations: each man checked his own equipment for any problems – tightening a loose buckle or closing an opened pocket that might snag on something in the train, holding the assault team up. All the men's personal belongings had already been placed in 'diddy' bags, and the attack platoon members would only carry their weapons, ammunition and explosives. Finally, each man checked his oppo, before being given the once over by the platoon sergeant and officer. Then it was time to move out.

The whole train shuddered with the vibration produced by the jets

Zero hour minus 30 minutes. The attack platoon made its way to the pre-arranged assault positions around the carriages, moving warily to preserve the element of surprise that was so crucial to the success of the operation. Three members of the platoon placed explosive charges on the train doors while other men placed small scaling ladders against the carriages. Five minutes before zero hour, all the marines and the RijksPolitsie in the area donned headphones and waited for the diversions to take place. A few miles from the train, Dutch Air Force F-104 Starfighters began their approach to the target.

Zero hour. The Starfighters came screaming in at roof-top level, and as they swept low over the train the pilots turned on the fighters' afterburners. The effect was instantaneous and, for those in the carriages, extremely terrifying. The whole train shuddered with the vibration produced by the jets; the hostages threw themselves to the floor. The South Moluccans had no time to recover their shattered senses, before the charges in front of the train were exploded by the Dutch SBS. Simultaneously, the covering platoon unleashed a blistering fusillade on the carriages that held several of the hijackers, carefully avoiding the second and fourth cars where most of the hostages were concentrated. To the South Moluccans, already confused and dazed by the pyrotechnics, it seemed that the authorities were trying to kill everyone on the train. However, none of the hostages was hit by the fire from outside.

Seconds later, the carriage doors had been blown off their hinges and the attack platoon was pouring into the train. Only one command was heard above the din: 'Get down'. All the hostages obeyed, except

Main picture: Surrounded by low-lying countryside, the hijacked train at De Punt (bottom) was a tough nut to crack. The authorities responded by throwing a tight cordon round the immediate area (right and below right), but as talks with the South Moluccans broke down, an anti-terrorist squad of the Dutch Marines was sent in (below).

for two unfortunate individuals who were tragically killed during the ensuing firefight. The marines were well prepared to identify the South Moluccans. They had spent many hours studying photographs of the hijackers and, if no picture existed, accurate photofits and artist's impressions were used.

The close-quarter drill used by the marines to deal with the hijackers was simple but highly effective. When they went into the carriages, they made sure that they were all facing in the same direction. Therefore, someone holding a weapon and facing in

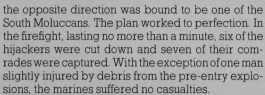

the opposite direction was bound to be one of the South Moluccans. The plan worked to perfection. In the firefight, lasting no more than a minute, six of the hijackers were cut down and seven of their comrades were captured. With the exception of one man slightly injured by debris from the pre-entry explosions, the marines suffered no casualties.

Zero hour plus five minutes. The train was secured and the hostages' ordeal was finally over. However, the marines faced weeks of post-attack debriefings. Whatever the authorities said in praise of the marines, they saw the mission as only a qualified success; most of the hostages had been saved, but two had died in the exchange of fire. Their aim had been to rescue all the hostages unharmed, even though they realised that their task was almost impossible. However, to recapture a train standing in the open, with little or no cover available and at the cost of only two hostages, was a remarkable act that reflected the courage and professionalism of the Royal Netherlands Marine Corps.

THE AUTHOR Nigel Foster was a member of the British Army Intelligence Corps. Following training at Lympstone, he was attached to 3 Commando Brigade, and was one of the unfortunates who spent four months on HMS *Bulwark* following the independence of Aden.

TUNNELS OF DEATH

Deep in an amazing complex of hand-dug tunnels, Viet Cong guerrillas lived, fought and died as they took on the most powerful military nation on earth

NAM THUAN, Communist Party secretary and political commissar of the South Vietnamese village of Phu My Hung and its six small hamlets, squinted into the early morning sun of an August day in 1968 and counted 13 M113 armoured personnel carriers rattling towards his command area. The Americans were, if nothing else, predictable. They usually came after breakfast, they came in great force, and they came with great noise.

Thuan's 'platoon' had been much depleted during the Tet Offensive mounted earlier that year, when the communists had launched their massive attacks on Saigon and other main cities in the South. The platoon now comprised only a good deputy commander and a couple of village farm boys. Thuan knew that once the Americans burst through to Phu My Hung, then a crucial and totally secret underground tunnel complex would be exposed. His job was to lure that small American strike force away in a long and diversionary action. Knowing the tunnels of Cu Chi better than he knew his wife's face, Thuan deliberately allowed the lead M113 to spot him.

He ran into open view, and slipped quickly through a trapdoor in the earth, and waited three feet below the surface. Seconds earlier, he had been outnumbered by men and machines. But down here, in the still, cool tunnel complex, Thuan had chosen his killing ground. As he waited, checking and re-checking his AK-47 automatic rifle, he silently prayed that, once again, the tunnels would save him and his village.

The tunnels of Cu Chi were an incredible hand-dug network of complex subterranean passages that

Right: Nam Thuan, whose cunning and bravery in a diversionary action saved the tunnels at Phu My Hung. Far left: Tunnellers at work. Above left: American GIs survey a piece of Viet Cong handiwork – a deadly punji-stake trap. Below left: A Viet Cong medical team. The tunnel complexes housed hospitals but shortages of medical supplies meant that many major operations, such as amputations, were performed without the luxury of anaesthetics. Below: Vo Mi Tho, a veteran female guerrilla, whose father had fought with the Viet Minh against the French.

GUERRILLAS

Communist forces in the first Indochina War against France were known as the Viet Minh (the Vietnamese people), though by 1959 they were designated Viet Cong (Vietnamese communists) to distinguish them from the people of North Vietnam. These rural, village-based guerrillas could draw upon almost 30 years' experience in unconventional warfare. Without heavy weapons or aircraft, they used their own battle skills – in particular a cunning, ruthless hit-and-run campaign, linked to political education of the people.
Within four years VC armed

strength rose from 5000 to about 40,000 by 1963. Their aim was to bring down the South Vietnamese government and inflict as many American casualties as they could. The only way to achieve this was to operate, literally, 'underground' in a massive tunnel network, left over from the war against the French. The system was expanded during the 1960s to cover a large part of South Vietnam.
Between 1965 and 1968, the Viet Cong controlled the villages and hamlets surrounding the towns, reaching the peak of their strength at the launch of the Tet Offensive.

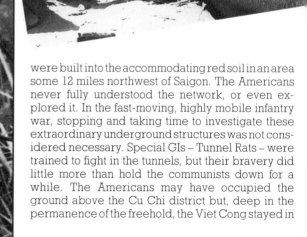

were built into the accommodating red soil in an area some 12 miles northwest of Saigon. The Americans never fully understood the network, or even explored it. In the fast-moving, highly mobile infantry war, stopping and taking time to investigate these extraordinary underground structures was not considered necessary. Special GIs – Tunnel Rats – were trained to fight in the tunnels, but their bravery did little more than hold the communists down for a while. The Americans may have occupied the ground above the Cu Chi district but, deep in the permanence of the freehold, the Viet Cong stayed in

Above: Viet Cong fighters risk instant annihilation as they dissect a dud US bomb to extract its high-explosive contents. Left: Slipping stealthily from their tunnel bases, the VC laid deadly ambushes for American patrols, and launched hit-and-run mortar raids against enemy fuel dumps (top). Top right: Captain Nguyen Thanh Linh, commander of the VC Cu Chi Battalion and expert in the tactics of tunnel warfare.

their tunnels, fighting a war against the most powerful military nation on earth.

The 200-mile underground network was created almost by accident. After World War II, the French returned to Indochina as colonial masters. The Cu Chi district had a long tradition of nationalism, and the French dealt harshly with the insurgent Viet Minh (as the predecessors of the Viet Cong were then called). Mobile guillotines were sent to Cu Chi to execute nationalists and, after a time, the Vietnamese found it prudent to dig small tunnels in the soil and hide from the French police. After the defeat of the French in 1954, a small but useful tunnel infrastructure remained. When the Americans arrived in strength a decade later, the North Vietnamese communist leader Ho Chi Minh ordered the cadres in the South to expand the tunnel system dramatically. Whole villages – men, women and children – dug by hand.

He crouched, waiting for the Americans to dismount from their M113s and inspect the trapdoor

Mai Chi Tho (today the mayor of Ho Chi Minh City, which was once Saigon), together with leaders from the North, planned the Saigon end of the Tet Offensive from the Phu My Hung tunnel that Nam Thuan was to protect so ruthlessly a little later. Captain Nguyen Thanh Linh, now of the People's Army of the Socialist Republic of Vietnam, spent five years commanding his sector of the tunnels. At the 7th Military Headquarters in Ho Chi Minh City he briefed us: 'The earth was perfect for making tunnels,' he explained. 'It is sticky and doesn't crumble. The whole Cu Chi

area is 65ft above sea level, and for 20ft down we knew there was no water.' The dry laterite clay, malleable during the monsoon but like brick in the hot seasons, was ideal.

Party instructions on how to dig, and the precise tunnel dimensions, actually fell into American hands during the war, but the significance of the papers was never fully appreciated. It was a magnificent, finely engineered system with astonishing 'rooms', kitchens, large chambers and hospitals, in which life continued for the Viet Cong. The key to protecting this whole series of subterranean hamlets was the security of camouflage on the outside. The tunnel entrance trapdoors were so finely engineered as to be invisible to the average Western eye. The authors have stood in a small clearing in the jungle, knowing for sure that there was a trapdoor at their feet, but still unable to spot it.

Around major tunnel entrances, the VC dug special 'spider holes' – one or two-man fighting pits, wonderfully camouflaged – allowing ace snipers to keep whole American squads at bay long enough for tunnels to be cleared if they were discovered. Even primitive booby-trapped pits were used to deflect American attention from coming too close to a tunnel entrance. Punji stakes, bamboo spikes smeared with excrement, were placed at the bottom of pits for the unwary and unlucky GI; special VC-made imitations of the terrible American Claymore anti-personnel mine were also used to blast GIs who wandered too near a tunnel entrance.

These were just some of the cards Nam Thuan held as he crouched, waiting for the Americans to dismount from their M113s and inspect the trapdoor tunnel entrance, which he had deliberately exposed to lure them in. His deputy commander lay well hidden, above ground, about 120ft away at another tunnel trapdoor. At the bottom of Thuan's short, three-foot shaft, and almost at right angles, began a 60ft communication tunnel. He lay comfortably at the junction considering his environment, letting his eyes suck in the available light. At the end of the 60ft tunnel was a similar short shaft leading up to the camouflaged trapdoor, where his number two lay hidden. It was vital to Thuan that this trapdoor was never found by the Americans. He crawled carefully into a small sleeping-alcove, dug a few feet into the wall of the tunnel. And there he waited.

Suddenly, he heard a muffled explosion, felt the blast, and blinked as a sudden beam of dust-filled sunlight pierced the shaft. The Americans had blown the trapdoor clean away. Good. It was what he wanted. The whole column would delay while the tunnel was explored. Thuan was the bait – Phu My Hung, with luck, would remain ignored.

Thuan leaned out of his alcove and, using the light from the tunnel entrance, shot the soldier twice

Nearly an hour passed before he heard the clap and whirr of the American helicopter. Thuan assumed the Americans were bringing in their special tunnel soldiers, the Rats. Even so, he felt no fear. Only one man could drop into the shaft at a time... and Thuan knew what to do to him.

The wiry little VC could not conceive of failure. He had already been awarded one victory medal, second class. He was about to earn another. His dream was to leave the small hamlet defence unit, and join the Regional Forces to take the offensive.

A small earth-fall from the now exposed tunnel entrance warned Thuan that the first GI was coming down. The American could only descend feet first, then wriggle awkwardly into the longer tunnel where Thuan lay hidden. In the past, as a GI's feet had touched the bottom, Thuan had stabbed the soldier in the groin with his bayonet. This time, as the green and black jungle boots came into view, Thuan leaned out of his alcove and, using the light from the tunnel entrance, shot the soldier twice in the lower body.

Above ground, the Americans were now in trouble. They couldn't drop grenades down the shaft because their mortally wounded comrade was jamming the hole – anyway, he might still be alive. But his body blocked the narrow shaft, preventing other soldiers giving chase. Thuan guessed it would take the Americans at least 30 minutes to slip ropes under

Below: An American soldier pumps pistol rounds into the mouth of a VC tunnel during a patrol in the War Zones north of Saigon. While the Americans fielded a special unit of tunnel warriors – the Tunnel Rats – going in to winkle out VC in the dark, confined passages of their tunnel bases was not an assignment the average GI relished. Far right: Nguyen Van Danh, an expert in mine laying, prepares another explosive surprise for an unwary American patrol. His expertise, however, had not come easily and the majority of his right hand was blown off during an abortive mine-laying mission earlier in the war.

Cu Chi District
1967-8

The tunnels of Cu Chi were the heart of a network running from Cambodia, through the Viet Cong-dominated countryside that became known as the Iron Land, and on to Saigon. At the height of the struggle for South Vietnam, the tunnel-dwelling Viet Cong guerrillas fought a bitter war of insurgency against their US and South Vietnamese enemies.

CAMBODIA

SOUTH VIETNAM

Cu Chi
Saigon

SOUTH CHINA SEA

Key
★ ARVN/US Bases
✦ Viet Cong headquarters
— Viet Cong tunnels

Phu My Hung

Song Saigon

Phu Hoa dong

Trung An

Cu Chi

To Saigon

Tan Phu Trung

On the night of 31 January 1968, the combined forces of the Viet Cong (VC) and the North Vietnamese Army (NVA) – involving a total of some 84,000 people – launched the Tet Offensive against targets in no less than five major cities, 36 provincial capitals, 64 district capitals and 50 villages throughout South Vietnam. The declared aims of the offensive were: to bring the South to such a state of turmoil that the government would collapse; to cause the South Vietnamese Army to collapse; and finally to undermine US determination to continue the war.

The operation was timed to coincide with the Tet (Vietnamese new year) national holiday celebrations, when the South Vietnamese would be least prepared to react effectively to the onslaught. The Viet Cong used the holiday to infiltrate into the towns and cities in the guise of holidaymakers, smuggling their weapons in with them.

The offensive opened with rocket and mortar attacks and fierce fighting followed these bombardments as the VC and NVA units struggled to establish themselves in the population centres. However, the tide soon turned against the communists as South Vietnamese and US troops re-took town after town, and by the end of the first week's fighting most cities and towns had been re-secured. During the offensive the NVA and VC took very heavy casualties and the number of people killed during the first two weeks of Tet has been estimated at 30,000. The great majority of casualties were among the VC units and after Tet the VC were so weakened militarily that they were never again able to function effectively. But, despite these losses, the war continued and Vietnam would see a further seven years of fighting before the South finally succumbed.

the dying man's arms and lift him out. The Americans' concern for their dead and wounded remained a constant source of bewilderment and relief to the communist soldiers.

Thuan's next fighting position was a second shaft, some four feet deep, which connected the first communication tunnel with a second, lower one. There was a trapdoor at the top of the second shaft, but he needed to remove it for his plan to succeed. He prayed the Americans would not be using gas at this stage to flush him out. They weren't. They fired ahead as they approached, but the bullets thudded harmlessly into the clay around him, sending little splinters of earth into the shaft in which he crouched …waiting. The GIs were even obligingly using flashlights as they advanced. They might as well have been using loudspeakers to announce the moment they came into range. Timing was now critical. The moment the Tunnel Rats saw the open shaft ahead, they would hurl a grenade into it, and Thuan would be blown to pieces. He heard a shout in a foreign voice. There was a lull in the firing. This was the moment. He thrust his head and shoulders out.

Flashlights blinded him, but all he had to do was fire the first clip from the AK-47. The noise was shattering. He loaded the second clip by touch, and fired that too. The tunnel exploded in a roar of noise, orange light and the screams of the dying. Then he ducked back into his shaft, replaced the trapdoor,

and crawled away. Later, his deputy commander reported that three bodies had been hauled out of the tunnel by the Americans. He was to kill at least three more the next day by entombing them in a shaft, then dropping a grenade on them.

Some 14 months later, Nam Thuan was invited to join the regular forces as an officer. He became fully responsible for the defence of six hamlets in the Phu An village complex. Some years later, on 28 March 1975, Thuan was with the forces who raised the flag of the Communist National Liberation Front over the town of Cu Chi. He is today a major in the People's Army of Vietnam.

Despite long and bloody operations to destroy the tunnels – particularly in the notorious Iron Triangle area – the Americans remained unsuccessful in their mission, and strangely uncomfortable with this unique kind of warfare. The tunnels and the Viet Cong inside them helped to pin down the best part of two divisions around Saigon at one time or another. 'The tunnels were a thorn stabbing in the eye of the enemy,' said Mai Chi Tho. They also became a symbol of the communists' determination to cling to their land and to maintain a presence among the uncommitted villagers, whose hearts and minds were more impressed with the resilience of their countrymen than the speeches of American presidents.

When American comedian Bob Hope played to the GIs at Cu Chi base, the Viet Cong organised their own subterranean entertainment for their soldiers in the tunnels. A communist playwright, Pham Sang, gave recitals and produced political plays on small stages deep inside the tunnels. Even in late 1968, women were giving birth inside the tunnels, weddings were being celebrated and whole companies of regular North Vietnamese soldiers were being stealthily guided through the network to the very

Viet Cong tunnel fighter, Vietnam 1967

This soldier wears the VC black 'pyjama' trousers and a North Vietnamese Army khaki service shirt. On his feet he wears 'Ho Chi Minh' sandals, constructed from truck tyres. He carries a Chinese Type 56 7.62mm assault rifle and three spare magazines in webbing pouches.

Below left: Smiling for the cameras of a Hanoi news agency, VC guerrillas set out on patrol.

gates of Saigon. And one major operation launched from the Tunnels – the Tet Offensive – was to signal the beginning of the end of American military involvement in Vietnam. It was also the turning point for the existence of the tunnels of Cu Chi. During a pause from bombing in North Vietnam, the weapon the VC tunnellers feared most was turned against the underground network. The huge B-52s rained high-explosive bombs onto the tunnels and many destroyed entire tunnel entrances before exploding.

By 1969 and early 1970, the tunnels had suffered enough damage to render their continued military usefulness dubious. But the character of the war was to change anyway. The Americans would leave, the war would become more conventional between the North Vietnamese Army and the ARVN soldiers of President Thieu in the South. By 1975, NVA tanks would be ploughing up the president's neat flower beds in the gardens of his palace in Saigon.

As many as 12,000 lives were lost fighting the tunnel war. Few of the original VC tunnel guerrillas survived, and only a handful of GI Tunnel Rats lived to tell their tales. The former underground headquarters at Phu My Hung are now maintained as a memorial to the tunnel war. The old conference chambers and the twisting tunnels that linked them have been carefully preserved around the village. Today it is a quiet place. A caretaker keeps things tidy. There is a visitors' book containing polite expressions of amazement from visiting delegations. Inside the tunnels, singing armies of mosquitoes and legions of fire-ants now dominate the still menacing blackness. Away from the old headquarters, the network lies in decay as tunnel entrances crumble and once-secret trapdoors become welded to the jungle floor through neglect. The Vietnamese young will tell you that hi-tech weaponry has meant the end of tunnel warfare. But the old are not so sure. They keep the maps safely stored – against another war at another time . . . but in the same place.

THE AUTHORS Tom Mangold and John Penycate revealed the story of the Tunnel Rats in their best-selling book *The Tunnels of Cu Chi*, to be republished in paperback by Pan Books this year. Both men are journalists with BBC1's *Panorama*.

The hardened professionals of the French Foreign Legion took on the dedicated revolutionary fighters of the Viet Minh in a desperate battle on the Cao Bang Ridge in 1950

FOR THE TWO companies of the 1st Battalion, 3ᵉ Régiment Etranger d'Infanterie (3 REI – 3rd Foreign Legion Infantry Regiment) manning the isolated, mud and bamboo fort at Phu Tong Hoa on a branch road of Route Coloniale 4 (RC4) in Tonkin, 25 July 1948 was drawing to a close like any other day. Sergeant Pierre Guillemaud, the quartermaster, and his Belgian friend, Corporal Pierre Polain, were cursing their ill-luck after an unsuccessful fishing trip to a nearby stream, and Captain Henri Cardinal, the post's commander, was swapping bawdy stories and rice wine, the potent local drink, with a few friends. The bored sentries in the post's four blockhouses had signalled the usual 'all quiet' to their comrades at Bac Kan a few kilometres to the south. It had, it seemed, been a normal day, except that, in the mist and downpour, no-one had spotted the 4000 Viet Minh guerrillas creeping into the scrub-covered hills surrounding the fort, or seen the sleek barrels of their artillery sited some 850m away.

At 1930 hours Guillemaud and Polain were enjoying the last of their MICs, the rough local cigarettes. They never finished their smoke. With an ear-splitting roar, four 75mm shells crashed into the main gate which disintegrated as other rounds burst on the barrack rooms, the captain's office, the radio hut and the cookhouse. Over 30 struck home before the hard-pressed Legionnaires silenced the enemy artillery with their mortars and a 37mm gun. The companies, however, were in a dangerous position; two blockhouses were gone, huge gaps had been blown in the perimeter fence and Cardinal lay dying in the wreckage of his command post. The remaining men of the garrison prepared themselves, 'lemons' (grenades) were primed and ammunition distributed. Then the storm broke in the form of a full frontal, human-wave assault launched with blood-curdling ferocity and intensity.

RIDGE OF DEATH

Below: The shattered remains of a French supply column after a Viet Minh guerrilla attack. Below, inset: A French Legionnaire prepares to go into action.

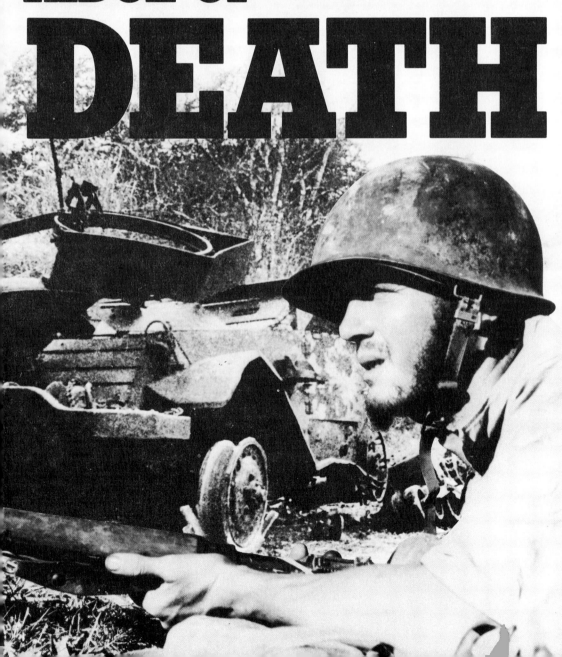

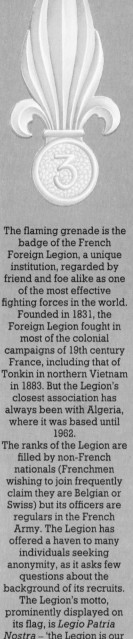

The flaming grenade is the badge of the French Foreign Legion, a unique institution, regarded by friend and foe alike as one of the most effective fighting forces in the world. Founded in 1831, the Foreign Legion fought in most of the colonial campaigns of 19th century France, including that of Tonkin in northern Vietnam in 1883. But the Legion's closest association has always been with Algeria, where it was based until 1962.

The ranks of the Legion are filled by non-French nationals (Frenchmen wishing to join frequently claim they are Belgian or Swiss) but its officers are regulars in the French Army. The Legion has offered a haven to many individuals seeking anonymity, as it asks few questions about the background of its recruits. The Legion's motto, prominently displayed on its flag, is *Legio Patria Nostra* – 'the Legion is our homeland'. And throughout its history, the men of the French Foreign Legion have been proud to fight and to die for their adopted military society. There were rarely less than 30,000 Legion troops in Indochina from 1946 to 1954 and units deployed included the 2nd, 3rd and 5th Legion Infantry Regiments, the 1st Legion Cavalry Regiment, the 13ᵉ Demi-Brigade de la Légion Etrangère (13 DBLE – 13th Foreign Legion Half-Brigade) – a force that had distinguished itself fighting on the Allied side after the fall of France in World War II – and two paratroop battalions, formed in 1948 and 1949.

Cao Bang Ridge
French Indochina, September – October 1950

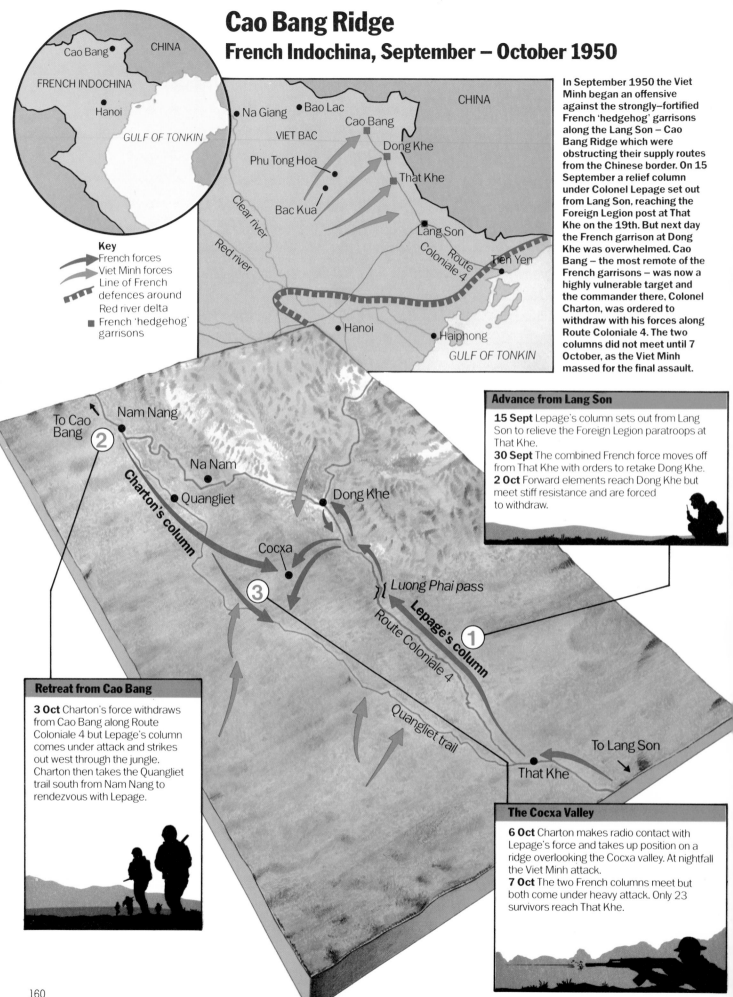

CHINA

Cao Bang

FRENCH INDOCHINA

Hanoi

GULF OF TONKIN

Key
→ French forces
→ Viet Minh forces
▰▰▰ Line of French defences around Red river delta
■ French 'hedgehog' garrisons

CHINA

Na Giang   Bao Lac

Cao Bang

VIET BAC

Dong Khe

Phu Tong Hoa

That Khe

Bac Kua

Lang Son

Clear river

Red river

Route Coloniale 4

Tien Yen

Hanoi

Haiphong

GULF OF TONKIN

In September 1950 the Viet Minh began an offensive against the strongly–fortified French 'hedgehog' garrisons along the Lang Son – Cao Bang Ridge which were obstructing their supply routes from the Chinese border. On 15 September a relief column under Colonel Lepage set out from Lang Son, reaching the Foreign Legion post at That Khe on the 19th. But next day the French garrison at Dong Khe was overwhelmed. Cao Bang – the most remote of the French garrisons – was now a highly vulnerable target and the commander there, Colonel Charton, was ordered to withdraw with his forces along Route Coloniale 4. The two columns did not meet until 7 October, as the Viet Minh massed for the final assault.

To Cao Bang

Nam Nang

②

Na Nam

Quangliet

Charton's column

Dong Khe

Cocxa

③

Luong Phai pass

Lepage's column

Route Coloniale 4

Quangliet trail

To Lang Son

That Khe

①

Advance from Lang Son

15 Sept Lepage's column sets out from Lang Son to relieve the Foreign Legion paratroops at That Khe.
30 Sept The combined French force moves off from That Khe with orders to retake Dong Khe.
2 Oct Forward elements reach Dong Khe but meet stiff resistance and are forced to withdraw.

Retreat from Cao Bang

3 Oct Charton's force withdraws from Cao Bang along Route Coloniale 4 but Lepage's column comes under attack and strikes out west through the jungle. Charton then takes the Quangliet trail south from Nam Nang to rendezvous with Lepage.

The Cocxa Valley

6 Oct Charton makes radio contact with Lepage's force and takes up position on a ridge overlooking the Cocxa valley. At nightfall the Viet Minh attack.
7 Oct The two French columns meet but both come under heavy attack. Only 23 survivors reach That Khe.

Five trumpet calls rent the night air, and within seconds hordes of screaming Viet Minh surged through the holes in the bamboo fence, and then flung themselves at the rock-like defenders of Blockhouse Three. Guillemaud lobbed grenades at the enemy while Corporal Polain, a sodden cigarette still in his mouth, stood over the body of a fallen Legionnaire and clubbed at the enemy with an empty rifle. A guerrilla crept behind the corporal and thrust a bayonet deep into his heart, but Polain's sacrifice was not in vain. The determined defence of the blockhouse won the day. By 2200 the Viet Minh onslaught seemed to be losing its momentum and Cardinal's replacement, 2nd Lieutenant Bevalot, ordered a counter-attack. Step by step, the garrison cleared the blockhouses, the breaches and compound. One of the Legionnaires remembered the French charge:

'From this moment the balance of the combat began to swing in our favour. Sergeants Andry and Fissler with three Legionnaires advanced, firing their automatics from the hip at point-blank range, and cleared the central building. Corporal Camilleri and two Legionnaires crawled through a breach and slaughtered the Viets who had occupied the north-west bastion. By 2300 the post was entirely in our hands and enemy trumpets could be heard sounding the retreat.

'A bloody dawn broke on the 26th. Within the walls lay over 40 Viet Minh dead; outside we counted more than 200. We had lost two officers (Cardinal and his second-in-command Charlotton had both died of their wounds) and 20 Legionnaires killed and 25 wounded.'

The Viet Minh military commander, Vo Nguyen Giap, was greatly dismayed by his men's failure at Phu Tong Hoa and set about making plans for a renewal of his offensive against the scattered French garrisons along RC4. By late 1950 his army had been expanded to 20,000 men, had acquired a firepower that was to shock the French, and his men, hidden along RC4 between Cao Bang and Lang Son, were ready for action.

'La route de la mort' (The road of death) with its jungle and scrub was ideal ambush country

Giap was presented with an ideal opportunity to blood his new army in early September after the French had decided that RC4 was too dangerous to hold, and that all the forts from Cao Bang to Lang Son had to be abandoned. The Viet Minh had been making life very uncomfortable for the garrisons in northern Tonkin and RC4, 'La route de la mort' (The road of death), with its jungle and scrub, its crags and gorges and 500 bends were ideal ambush country. Indeed, the road was so dangerous that the Legion was losing more men in guerrilla attacks on its supply columns than on its posts. In posts like Dong Khe and That Khe, officers and men sensed that Giap was preparing to pick them off. Phu Tong Hoa had been a bloody dress-rehearsal for the horror to come.

General Carpentier, the French commander, recognised that the most hazardous part of the plan would be the evacuation of the Cao Bang garrison, consisting of the 3rd Battalion, 3 REI and a battalion of Moroccans, and decided that a mobile group, code-named 'Bayard', should move from Lang Son to help in the withdrawal. This force, consisting of two Moroccan battalions and the 1er Bataillon Etranger

Parachutiste (1 BEP – 1st Foreign Legion Parachute Battalion) under Major Ségretain, and commanded by a gunner officer, Colonel Lepage, was to rendezvous with the Cao Bang column at Kilomètre 28, a point on RC4 dominated by the fortified village of Dong Khe, and was then to withdraw, gathering the remaining garrisons on the way back to Lang Son. Colonel Charton of 3 REI, the commander of the Cao Bang garrison, was far from happy. He foresaw that the retreat might well turn into a death march. In addition, apart from the men of 1 BEP, he had little faith in the fighting abilities of Lepage's group, and even less faith in Lepage, whom he described as a 'tired old man just hanging on until retirement'. Despite Charton's fears, the operation, timed to begin on 3 October, was given the go-ahead.

Although all the orders issued were supposedly top secret, the French plan was soon common knowledge and Giap was able to pre-empt the move by attacking Dong Khe. This outpost was held by two companies of the 2nd Battalion, 3 REI, which had recently taken over from a Moroccan battalion. An officer of one of the companies, who was badly wounded during the action and held prisoner for the next four years, remembered the Viet Minh attack:

'At midnight (18/19 September) and again at 0400 hours, the southern outpost had signalled movement and the sound of digging. Patrols were sent out to investigate and both bumped into the enemy, just as their guns and mortars opened up on the citadel, the Dubouchet quarter and the eastern outpost. One hour later the volume of fire was increasing, and we received a message from the eastern post stating that a shell had burst within the walls, destroying a machine gun and killing all the crew. Towards 1000 trumpets and bugles sounded, and the Viets came swarming down the adjacent hills. Some were carrying scaling ladders, and as they attacked, loudspeakers called on us in Arabic to surrender. The Viets had not realised that the Moroccans had been withdrawn.

'By 1100 the eastern post was in flames. Of the garrison of 11, only one man managed to fight his way back to the citadel. At 1800 hours, the citadel and northern post became the main targets for the enemy artillery. Then the northern post, held by 12 Legionnaires, fell and, an hour later, the enemy infantry, attacking in waves, fought their way through a breach in the northwest wall of the citadel and penetrated to the kitchen. A bloody hand-to-hand encounter followed. Legionnaires, employed on cookhouse fatigues, fought with any weapon they could lay their hands on. The sergeant cook and a Viet were found locked in a death embrace; the former pierced by a bayonet, the latter with a butcher's knife plunged in his heart.'

Fighting continued all night. By dawn the defenders had lost 120 out of their total strength of 230. They could ill-afford the loss as the attacking force consisted of 16 battalions.

By the next day it seemed that the Viet Minh were settling down for a siege, but in the late afternoon they resumed their attacks. By nightfall only the central bastion, the citadel, was holding out. During most of the night, the Viet Minh limited themselves to bombarding the remaining French positions. By dawn, however, they were ready to mount their final assault. A French officer saw the Legionnaires' preparations to meet the onslaught:

'Three lieutenants could be seen giving their last

Few of the Legionnaires taken prisoner by the Viet Minh near Dong Khe survived the horrors of a brutal captivity. The 700 who did were freed on 2 September 1954 after nearly four years in guerrilla hands. Many were suffering from the effects of torture and severe malnutrition. Both of the French commanders, Lepage (below) and Charton (bottom), were released, but Charton later died from the wounds he had received in the retreat from Cao Bang.

THE LEGION IN INDOCHINA

The Legion's association with Indochina had begun in 1883, when it provided two battalions for the conquest of Tonkin in northern Vietnam, and battalions of the Legion were kept on permanent station in Vietnam. The Japanese took overall control of Indochina during World War II, and, in March 1945, launched a surprise attack on French garrisons. The only French forces to escape were those members of the 5ᵉ Régiment Etranger d'Infanterie (5 REI – 5th Foreign Legion Infantry Regiment) who made a forced march into southern China. After the surrender of the Japanese in September 1945, the French managed to take back control of the south of Vietnam, but in the north, Vietnamese communist nationalists under Ho Chi Minh resisted and a guerrilla war on a large scale was in progress by 1947. The brunt of the fighting was borne by the regulars of the Legion and troops from France's African colonies. The Viet Minh guerrillas were based in the mountains known as the 'Viet Bac', and by 1949 the French had taken the decision to evacuate all their posts in the mountainous hinterland of northern Vietnam, preferring to concentrate on holding the fertile, heavily populated Red River Delta. All they retained was a string of forts along the Cao Bang-Lang Son ridge, which they believed would inhibit the flow of supplies from China (which had become a communist state in 1949) to the guerrillas. By 1950, however, the Viet Minh had received much equipment from the Chinese, and were preparing to assault the lonely outposts on the ridge.

orders to some 30 Legionnaires, the only survivors of eight platoons. The last case of grenades was opened, and its contents were distributed. Not 20m away, the Viets rose up for the final assault; their blackened faces under bamboo-camouflaged helmets, showed only intense hatred. On 20 September, eternal night descended on the defenders of Dong Khe. A citadel, in flames and strewn with corpses, had succumbed to an inexhaustible mass of men.'

The Legion had lost 85 men killed and 140 captured in the battle. Only five men broke through the enemy to reach That Khe and tell of the slaughter at Dong Khe. Giap had sacrificed 800 men, but he had finally cut RC4 and isolated Cao Bang.

The loss of Dong Khe forced the French to revise their original plan for the evacuation of Cao Bang. On 30 September, Lepage, who had been kicking his heels at That Khe for ten days, was ordered to march on Dong Khe and recapture the fort and then link up with the Cao Bang garrison moving south along RC4. The order was suicidal. Lepage had under 2000 men

and was faced by a well-equipped and elusive foe. He set out, however, and by 2 October the column stood outside the objective. Lepage gave the order to attack; the paratroopers and Moroccans were detailed to hold the hills to the east of the target while the rest of his force attacked from the west. The Viet Minh were waiting and both columns ran into trouble. Worse was to follow; French aircraft signalled that a vast guerrilla army was massing in the hills around Lepage's position.

Lepage had walked into a hornet's nest and could expect no help from Charton who had been held up. Burdened with its guns and 15 trucks carrying the wounded, his column had covered only 17km on 3 October, the first day of its withdrawal from Cao Bang. That evening Charton was ordered to leave the road and strike off into the jungle along the rough Quangliet trail to rescue Lepage's column.

A radio message was received stating that all Charton's trucks, guns and stores were to be destroyed and that the move was to be completed within 24 hours. The column plunged into the jungle in

Below left: Living rough was no hardship for the Legionnaires in northern Tonkin. All had been taught to live off the land in basic training. Food was often prepared under the most primitive conditions. This man is cooking a meal on an open fire. Below centre: The retreat of Charton's force down the Quangliet trail was the most hazardous part of the evacuation plan. Here, a Legionnaire armed with a US M1A1 carbine is giving medical aid to a wounded comrade during the retreat from Cao Bang. Below right: Two German Legionnaires sighting a heavy mortar during Lepage's abortive assault on Dong Khe on 2 October 1950.

search of the trail and, although it was found fairly quickly, it was so over-grown that the column's progress was extremely slow.

It took Charton's battered column about three days to reach the area in which Lepage's men were fighting for their lives after they had moved westwards from Dong Khe. By 6 October Charton had established a base along a ridge overlooking the Cocxa Valley where Lepage's force was surrounded and pinned down by the Viet Minh who had taken possession of the high ground. Realising the serious nature of the situation, Charton requested permission to retreat to That Khe, but Lepage insisted he stay, and the fate of both columns was sealed. Charton had little time to organise his defences before the guerrillas attacked. The fighting was severe and protracted as the regimental history of 3 REI testifies:

'Barely had this order been carried out before a Viet regiment fell on our rearguard and occupied the heights dominating the road. If a single man was to survive the crests had to be retaken. Led by Major Forget, the battalion drove the enemy off before they had had time to dig in. A first counter-attack was repulsed, but a second evicted the Legionnaires from their more advanced positions. During the course of the morning the positions changed hands several times, but while the struggle raged the rest of the column progressed slowly.

'Yet another attack pushed the Legionnaires from the vital ridge. Once more the battalion, with Major Forget at its head, charged with the bayonet. The Viet Minh stood their ground, fighting it out hand-to-hand, but then at last wavered and fell back. Hit in the thigh, groin, chest and head, Major Forget's last words were *"Je meurs fier de mon bataillon"* ("I die proud of my battalion").'

The battle that had raged throughout the night was, however, only a foretaste of the vicious hand-to-hand combat to follow. At 0300 hours on 7 October the guerrillas launched a full-scale onslaught against Charton's dwindling forces along the ridge. The situation seemed hopeless; his men held only a small saddle of the ridge some 900m long, they were facing an enemy enjoying overwhelming strength and the scheduled link-up with Lepage had still to be made. It was time for desperate measures. Lepage called on the Legionnaires of 1 REP for a final effort: 'Break through whatever the cost. The fate of the column is in your hands.' There were no more than 450 Legionnaires left and they faced twenty times that number of Viet Minh holding the vital peaks between them and Charton's embattled column. The men had hardly slept in the last two days, but they were all prepared to make a final sacrifice. In a downpour that shrouded the jungle in an eerie mist and drenched their tattered uniforms, the men gathered their rifles and grenades, and set off to climb the ridge's limestone slopes. Suddenly, the Viet Minh opened up with smallarms fire and men began to fall, but the paratroopers pressed on:

'At three in the morning of 7 October the 2nd Company attacked the machine-gun nests barring their way. The attack was repulsed. The 1st Company then moved in from the right while the 3rd charged frontally. Then, as a boxer shakes his opponent by showering punches on him with both fists, the parachute battalion shattered the enemy strongpoints. After several hours of desperate combat, with both sides displaying the most supreme bravery, the Legionnaires realised that the Viets were softening. The paratroopers left alive made a final effort and the breach was opened.'

The cost, however, had been frightful. The paras had destroyed themselves in their attack and less than 100 men survived.

A bare 23 survivors reached That Khe. The withdrawal had cost the French 4000 killed

Charton had heard the fury of the desperate battle and at dawn on 7 October he saw a horde of frantic, ill-disciplined Moroccans, the only other survivors of Lepage's column, run into his own positions. Clearly there was nothing more to be done. Further resistance was pointless, and Charton ordered his men to split into small groups and make for Lang Son via That Khe. Many fell prey to Viet Minh ambushes as they tried to escape and only 23 survivors, many little better than scarecrows, reached That Khe. The withdrawal 'with honour' along RC4 cost the French over 4000 killed.

The Legionnaires of 3 REI and 1 BEP had faced the might of the Viet Minh in the merciless jungle of northern Tonkin and, although their destruction testified to the valour with which they had fought, they could not compensate for the failure of their generals to grasp the initiative from Giap after his troops had stormed Dong Khe in September. The Legion has always held that there is honour in defeat, and the men who fell in October 1950 died with unswerving belief in this ideal.

THE AUTHOR Lieutenant-Colonel Patrick Turnbull commanded 'D' Force, Burma, during World War II. He has published numerous books including the *The Foreign Legion*.

HEROES
OF THE HORSESHOE

The Horseshoe, a ring of hills lying south of the Mareth Line in Tunisia, was one of the Grenadier Guards' toughest objectives in the campaign against the Afrika Korps

REPOSING IN THE chapel of the Grenadier Guards' depot in Purbright is a small, white stone cross dedicated to the memory of those soldiers who fell at the Battle of the Horseshoe in March 1943. An icon kept in the home of one of the oldest regiments of the British Army, it reminds new recruits of what it means to be a guardsman.

On 16 June 1942, the 6th Battalion, Grenadier Guards, commanded by Lieutenant-Colonel A.F.L. Clive, left England and sailed for North Africa to join the 201st Guards Brigade, under the command of Brigadier J.A. Gascoigne. After flying the British flag during a goodwill tour of Palestine and Syria, in February 1943 the battalion made an overland jour-

ney of 2200 miles to Medinine, in southern Tunisia, to join General Bernard Montgomery's Eighth Army. One month later, on 6 March 1943, the Battle of Medinine pushed Axis forces behind the Mareth Line and provided Montgomery with the opportunity to attack in force.

Although weakened and demoralised, the German and Italian forces under General Giovanni Messe remained formidable opponents, entrenched behind a system of fortifications stretching 22 miles from the Mediterranean to the Matmata Hills in the east. Concrete blockhouses and the Wadi Zigzaou afforded Messe's five Axis divisions a powerful block against an Allied assault.

Montgomery laid his plans carefully, deciding to initiate a preliminary battle that would enable the

Below: Photographed in dramatic silhouette, a 25-pounder gun crew add their shells to a large-scale rolling barrage in support of advancing infantry. The 25-pounder was an immensely valuable weapon in the Western Desert, being easily manoeuvrable and having a superior range to both German and American 105mm artillery of the period.

Above: General Bernard Montgomery (centre) speaks with an officer of the Grenadier Guards during a visit to their positions in Tunisia. To Montgomery's left is Lieutenant-Colonel A.F.L. Clive, MC, CO of the 6th Battalion and active participant in the Battle of the Horseshoe. Left: A roadside marker bearing names made famous during the Eighth Army's Desert Campaign. Bottom: Troops pick their way through the wire defences of the Mareth Line.

51st Highland Division to launch a frontal assault over the Wadi Zess without fear of being outflanked. Elements of the 90th Light Division of the Afrika Korps manned an outpost position south of the Mareth Line, in a horseshoe-shaped area of hills straddling the Gabes-Medinine road. The 201st Guards Brigade was therefore ordered to secure the area. Aerial reconnaissance had located no minefields in the vicinity. The assault was to be preceded by an artillery barrage.

After Montgomery had visited Gascoigne's Brigade Headquarters, he left with the words, 'When I give a party, it is a good party. And this is going to be a good party.' Unbeknown to the 6th Battalion, however, the enemy had been forewarned – Montgomery's party was to have a great many uninvited guests.

On 15 March, 24 hours before the attack, a patrol under Lieutenant J.K.W. Strang-Steel was surprised by the enemy and two guardsmen were taken prisoner. In addition, an artillery officer of the Highland Division was captured while carrying a map on which the successive lines of the barrage were clearly indicated. German intelligence reacted with

The 6th Battalion was raised at Caterham on 18 October 1941. The 3rd and 5th Battalions had been formed in 1940, and the formation of these three new units released the 1st 2nd and 4th Battalions for training in armoured warfare. Commanded by Lieutenant-Colonel A.F.L. Clive, the 6th Battalion was earmarked for early overseas deployment. The battalion left England on 16 June 1942, sailing to Suez before joining Brigadier J.A. Gascoigne's 201st Guards Brigade at Qatana, Syria. In November 1943 it toured northern Syria. Between 7 February and 6 March 1943, the battalion made an overland journey of 2200 miles through Syria, Palestine, Egypt and Libya.

Held in reserve by Gascoigne during the Battle of Medinine on 6 March, the battalion's debut in action was at the Battle of the Horseshoe on 16/17 March. The end of the Tunisian campaign was followed by a period of re-equipment and training at Bone in Algeria. In August 1943 the battalion moved back to Tripoli and one month later, on 9 September, it became the first Grenadier battalion to set foot on mainland Europe since Dunkirk, landing at Salerno in Italy as part of the Allied invasion force.

After fighting with distinction during the struggle for Bare Back Ridge at Camino, on 8 March 1944 the battalion was withdrawn from the line and reorganised – 17 officers and 400 men being assigned to the 5th Battalion, and 200 men sent to the Reinforcement Training Depot. The remaining officers and men returned to England, and on 4 December 1944 the 6th Battalion was disbanded.

Following the Battle of Alamein (23 October – 4 November 1942), General Montgomery's Eighth Army pursued Rommel's Afrika Korps through Tobruk, Benghazi and Tripoli. On 18 February 1943 Axis forces, comprising the German Fifth Army and the Italian First Army, retired behind the Mareth Line – a system of fortifications originally built by the French to check an Italian invasion of Tunisia.

On 6 March 1943, Rommel sent three panzer divisions to attack the Allied southwest flank at Medinine. However, his intentions became known to Allied intelligence. Anti-tank guns were sited in depth, and Rommel's forces were comprehensively defeated by the Coldstream and Scots Guards of the 201st Guards Brigade. Rommel left North Africa three days later and was replaced by General Giovanni Messe.

Assisted by a diversionary attack (Operation Wop) from the north by General George Patton's II Corps, US Army, Montgomery launched a three-phase attack on the Axis forces entrenched behind the Mareth Line. The first phase was a preliminary attack by the 201st Guards Brigade to clear elements of the German 90th Light Division from the high ground to the south of the Mareth Line.

This was codenamed Operation Walk. Although the Guards were forced to retreat, three days later Montgomery launched the second phase of his assault – with XXX Corps attacking in the centre, and the New Zealand Division, under General Bernard Freyberg, leading a flanking movement through the Matmata hills to El Hamma.

When the 15th Panzer Division repelled the frontal assault by XXX Corps, Montgomery switched the weight of the offensive over to Freyberg's sector. This strategy, known as 'Supercharge II' included a daylight air assault and a rolling artillery barrage. It forced the Axis forces to move troops to the right flank and, as a result, the British 51st Highland Division broke through the weakened Mareth Line on 27 March.

lightning speed, enabling elements of the 90th Light Division to deploy in strength along the line of the brigade's intended advance towards the Horseshoe.

At 1930 hours on 16 March, unaware of these developments, the brigade moved forward into the broken terrain of the front line; the Grenadier Guards were on the right, and the 3rd Battalion, Coldstream Guards, on the left. The moonlight enabled the Grenadiers to catch sight of their objective; a group of barren hills that lay dwarfed by the Mareth Mountains on the horizon.

Lieutenant-Colonel Clive ran over his plan of attack with his officers. He had designated three assault companies to attack to the right, centre and left of the Horseshoe – respectively, they were No.3, under Captain G.C. Gwyer; No.1, under Major Peter Evelyn; and No.4, under Major Thomas Butler. A consolidation group, comprising No.2 Company with scout platoons, medium machine guns, mortars and essential vehicles, was tasked with warding off any enemy counter-attack. Commanded by Major W.H. Kingsmill, this group would have to negotiate the high banks of the Wadi Zess before being able to join up with the three assault companies. Drill and discipline had made the men the best turned-out soldiers in North Africa and, with their khaki trousers sharply creased, they marched in high spirits through the olive groves that led to the Wadi Zess.

The Germans, hidden by an elaborate system of trenches, knew that they could afford to hold their fire

Hitting the ground at the start line, 500yds short of the wadi, the Grenadiers fixed bayonets and waited for the artillery to begin the bombardment. Clive passed through the ranks, murmuring words of encouragement to his men. At 2045, the still night air was shattered by the crash of guns as the 25-pounders of eight artillery regiments concentrated their rolling barrage on a 2000yd front. This was the

Below: Following the breaking of the Mareth Line on 27 March 1943, British troops watch from the Mareth fortifications as Axis forces move northwards to the Wadi Akarit.

signal to move and, encountering only sporadic enemy fire, the battalion advanced over open ground and crossed the wadi with few casualties.

While the battalion was re-forming, an advance patrol discovered a thin wire running parallel to the wadi about 50yds north of the assault companies. The preliminary reconnaissance and intelligence had been mistaken, for the enemy had been able to lay an extensive minefield as its first-line defence. The Germans, hidden by an elaborate system of trenches, knew that they could afford to hold their fire.

At first, blood-red tracer streaks passed harmlessly over the Grenadiers' heads as they advanced up the hillside. When a series of explosions violently rocked the ground in all directions, however, the assault companies realised the extent of their predicament. Casualties mounted, and increased further when the German mortar teams opened up. The guardsmen were caught in a trap. As deadly splinters from the mortar bombs sliced past them, they probed their way towards the German positions.

One officer later described this horrific detail:

'At one moment, I thought that an Indian-file method of crossing the minefields would be best, but when several men following closely behind me were blown up I revised my view. Dispersal should be every man's aim, so that no mine caused more than a single casualty.'

The German troops, watching and waiting in their trenches, were astounded by the bravery of the infantry advancing inexorably towards them. Captain Gwyer, despite being badly wounded, insisted on being carried forward to No.3 Company's objective. Lieutenant N.S.T. Margetson set out determinedly to silence a machine but he was never seen alive again.

Finally, one hour after moving off, the assault companies got through the minefield, only to discover another lying in their path. Clouds of dust, thrown high into the air by the artillery barrage, mortar fire and exploding mines, added to the confusion of battle. But the guardsmen were unswerving in their

The Battle of the Horseshoe
6th Battalion, Grenadier Guards, 16-17 March 1943

After the decisive defeat inflicted on the Afrika Korps at El Alamein in late 1942, the Eighth Army began a headlong drive to the gates of Tunisia that was finally brought to a halt by a series of Axis defences known as the Mareth Line. Although Operation Torch, the Allied landings in Algeria and Morocco during early November 1942, had forced the Axis forces to fight on two fronts, the Allies were still looking for a decisive battle. After a pause to regroup and re-organise, the Eighth Army began operations against the Mareth Line in March 1943. To test the quality of the German defences, a number of probing attacks were organised. On 16 March, the 6th Battalion, Grenadier Guards, was launched against the Horseshoe.

The Drive to Mareth, December 1942-February 1943

Gabès
Mareth
Medenine **16 Feb**
TUNISIA
Tripoli **23 Jan**
Coradini **21 Jan**
Homs **19 Jan**
Tarhuna
Misurata
19 Jan
MEDITERRANEAN SEA
Buerat
Sirte **21 Dec**
Nofilia
15 Jan
Marble Arch
L I B Y A

Key
→ Eighth Army's advance
----- Axis defence lines

The Mareth Line

Gabès
MEDITERRANEAN SEA
Axis withdrawal
TUNISIA
Mareth
marshes
Bene Zelten
Wadi Zigzaou
Wadi Zess
Eighth Army attacks
Horseshoe
Toujane
Matmata Hills
Medenine
Ksar el Hallouf

Into the Horseshoe

16 March

1930 The Grenadier Guards begin to move up to the Horseshoe.
2045 The attack gets under way. Preceded by an artillery barrage, the Guards negotiate two minefields and push on towards the enemy's front line.
2230 No. 3 Company takes Hill 139.

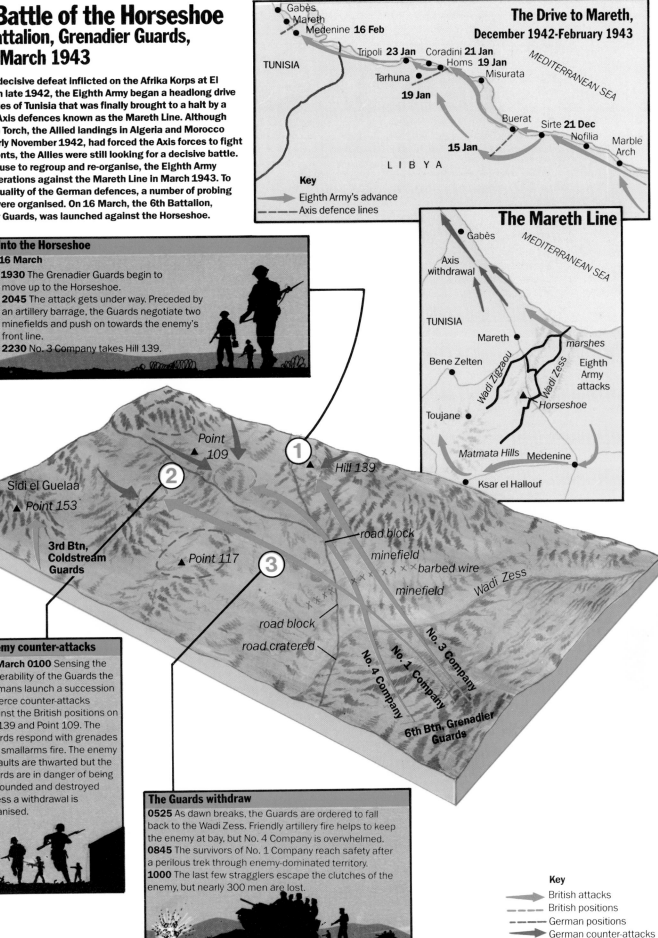

Point 109
Sidi el Guelaa
Point 153
3rd Btn, Coldstream Guards
Point 117
Hill 139
road block
minefield
barbed wire
minefield
Wadi Zess
road block
road cratered
No. 3 Company
No. 1 Company
No. 4 Company
6th Btn, Grenadier Guards

Enemy counter-attacks

17 March 0100 Sensing the vulnerability of the Guards the Germans launch a succession of fierce counter-attacks against the British positions on Hill 139 and Point 109. The Guards respond with grenades and smallarms fire. The enemy assaults are thwarted but the Guards are in danger of being surrounded and destroyed unless a withdrawal is organised.

The Guards withdraw

0525 As dawn breaks, the Guards are ordered to fall back to the Wadi Zess. Friendly artillery fire helps to keep the enemy at bay, but No. 4 Company is overwhelmed.
0845 The survivors of No. 1 Company reach safety after a perilous trek through enemy-dominated territory.
1000 The last few stragglers escape the clutches of the enemy, but nearly 300 men are lost.

Key
→ British attacks
----- British positions
----- German positions
→ German counter-attacks

determination to take the high ground. Their earlier high spirits had given way to an overwhelming sense of fury and the desire to lock horns with the enemy at close quarters.

At 2230, the sound of enemy gunfire on Hill 139 was silenced – No.3 Company had taken its objective. Minutes later, the company fired red and green Very lights to confirm its success to Lieutenant-Colonel Clive. By 2320, the other two companies had surged over the final ridge and swept the Germans from their trenches. Those who refused to surrender were bayonetted, and guardsmen ran along the line of the defences blowing out enemy strongpoints with a few well-placed grenades. The Very lights went up, and were greeted with rousing cheers from the consolidation group back on the start line.

As silence descended over the hills, the assault companies began to take stock – the cost of achieving their objectives had been high. No.4 Company had suffered 70 per cent casualties and was desperately short of ammunition. Major Evelyn's No.1 Company had been reduced to 35 men, and No.3 Company had been similarly devastated. Cut off from Battalion Headquarters and each other by faulty wireless sets, the three companies were dangerously isolated. As the moon began to sink, casting eerie shadows over the hills, the enemy launched their counter-attack.

Nos.1 and 4 Companies had been forced to switch objectives during the confusion of battle, and had unknowingly bypassed a strong concentration of German troops during their advance. The largest pocket lay on a hillock southeast of No.4 Company's position on Point 109. Just before midnight, having recovered from the artillery barrage, German forces reacted with a fierce broadside against the rear of the assault companies.

By now, enemy fire from Point 117 was threatening the forward companies, and it was also preventing the consolidation group from crossing the Wadi Zess and the minefields. Aware that their colleagues on the Horseshoe were in urgent need of reinforcement, sappers worked frantically to dig tracks through the wadi that would allow the passage of the

Below: Infantrymen storm a rocky promontory during General Freyberg's outflanking movement to Nofilia, one of the highlights of the war in Libya in late 1942. Attacking the Mareth Line in March 1943, General Montgomery sent Freyberg's New Zealand Corps on a wide detour around the Matmata Hills and north to the Tebega Gap, an act which forced the Germans to pull back from the Mareth Line to ward off the threat.

group's Bren-gun carriers and anti-tank guns. Several vehicles exploded under the constant mortar bombardment and heavy machine-gun fire, illuminating the surrounding area and presenting ready targets to the Germans.

Casualties were very heavy, but the officers and subalterns continued their attempts to find a way through for the scout platoons and heavy vehicles. The commander of the anti-tank company, Major A.J. Gordon, was seriously wounded but still managed to co-ordinate ammunition supplies to No.3 Company. Lieutenants C.E. Trimmer-Thompson and A.G. Buchanan, and Captains Goss and Allsopp, all fell victim to enemy mines.

Sensing their opponents' isolation, the Germans increased the ferocity of their attack. By 0100 hours on 17 March, the right flank of No.3 Company was coming under heavy fire, and the men of No.4 Company threw a continuous stream of grenades to ward off German soldiers who were creeping up on all sides. From the Advanced Battalion Headquarters, 500yds north of the wadi, Lieutenant-Colonel Clive gathered together a small patrol and made a sweep around the immediate area of the headquarters. A further recce, in the vicinity of No.3 Company, convinced Clive that the consolidation group had to find a way through. If it failed, the forward companies woud be forced to withdraw alone and in the face of withering enemy fire.

Enlisting the aid of a few Bren carriers that had survived, Wiggin set out across the minefield

Kingsmill formed a defensive flank on the southern side of the wadi to guard against attack from the northwest, and sent Lieutenant J.H. Wiggin on a mission to find No.1 Company. Enlisting the aid of a few Bren carriers that had survived the crossing to the enemy side of the wadi, Wiggin set out across the minefield. Threading its way past cutting machine-gun fire from Point 117, the patrol chanced upon a track running to the northwest between the positions of Nos.1 and 4 Companies. Remarkably, the track was clear of mines. Wiggin halted briefly on the slopes of Point 109, searching in vain for Major Evelyn and Major Butler. In fact, he was close, but not quite close enough – Evelyn had shouted for assistance, but the sound of his voice failed to carry above the noise of battle. Making a wide loop around the front of No.1 Company, Wiggin returned to Battalion Headquarters with the loss of only one carrier. A way through had been discovered, but the exact positions of the forward companies remained unknown.

Above: Troops undergo the ordeal of clearing mines from their path as the enemy pours fire down on them. A Bren gunner is laying down what covering fire he can as the men operate their primitive mine detectors. Above right: Their battledress characteristically well pressed, members of a Grenadier Guards battalion man a frontline slit trench with rifles and Thompson sub-machine guns.

Sergeant, Grenadier Guards, Tunisia 1943

This sergeant is wearing serge battledress with regimental shoulder title and sergeant's chevrons, a hessian-covered steel helmet, '37 pattern web anklets and ammunition boots. He is carrying a captured German 7.92mm MG34 machine gun, which was the first example of what is known today as a 'general purpose' machine gun. It was supplied with a bipod to serve as a squad's light automatic support weapon, and also a tripod to allow it to function in the sustained-fire mode of a medium machine gun. Fitted with a drum magazine it was also effective as an anti-aircraft weapon.

The time was 0430, the battle had been raging for eight hours and soon daylight would dispel the darkness, forcing the guardsmen to withdraw under point-blank fire from the German machine guns.

Despite many casualties, the consolidation group split into three parties and endeavoured to retrieve survivors from the battlefield. Lieutenant Vaughan led a platoon of four anti-tank guns, Sergeant-Major Dowling led an infantry platoon and Lieutenant Wiggin again braved enemy fire with the five carriers that remained under his command. Those wounded who could be reached were ferried back to a makeshift first-aid post where they were treated under a storm of mortar fire.

At 0525, just before the first streaks of dawn, Brigadier Gascoigne ordered a withdrawal to the southern side of the wadi. The artillery went into action again, sending up smoke bombs to cover the companies' retreat across open ground.

The acting commander of No.3 Company, Lieutenant T.G. Ridpath, had served as an inspriation to his men during their terrible ordeal. Three times during the night German soldiers had penetrated their perimeter, and three times Ridpath had sent them scurrying away. Ridpath was killed just before the withdrawal, and it was left to Sergeants Harrison and Delebecque to lead the remaining handful of men back to the wadi. Leaving the shelter of their trenches, the remnants of No.1 Company retraced their steps across the minefield and reached headquarters at 0845. Not a single officer had survived.

Major Butler received the order for No.4 Company to withdraw, but his unit was surrounded. Captured after a grenade exploded in his face, Butler later escaped to Allied lines and described his company's heroic last stand at Point 109:

'There is a very close and ever-tightening ring around our position and a great number of machine guns and automatics pour a continuous stream of bullets over our heads...I call for artillery fire all around our hill. This call is promptly answered, but doesn't appear to do a great deal of harm, in spite of shells bursting all around us...I decide that the time has come for us to fight our way out rather than be massacred in our trench-

es... A few of us make the attempt, only to be shot down within a few yards of our trench.'

Butler's company had penetrated into the heart of the German defences and, after bitter close-quarters fighting, the survivors were overwhelmed.

Throughout the night, communications with No. 1 Company had been only intermittent and, after the radio mast at headquarters was shot down, Gascoigne's order to withdraw could not reach the remaining 30 men under Major Evelyn's command. Spurred on by the capture of Point 109, German troops began to close in. Under fire from the northeast and southwest, Evelyn shouted the order for his company to re-form. He had taken the decision to withdraw, but looking around at the men Evelyn realised that such an order would be a severe blow to their morale. Showing great presence of mind, he informed the company that it was going to advance into the teeth of the enemy and take Point 117.

At that moment, four carriers crested a ridge and drove up to the ranks of the company. Lieutenant Wiggin had returned. Loading up, the carriers set off back down the track, running a gauntlet of fire from Points 109 and 117. Three of them completed the perilous journey but Wiggin's vehicle was set ablaze, forcing 14 guardsmen, including Wiggin and Evelyn, to dive into the shelter of a nearby ditch. Scattered in small groups, the men lay exposed to enemy fire while they awaited the return of the

Above left: King George VI inspects a guard of honour of members of the 6th Battalion, Grenadier Guards, after the Battle of the Horseshoe. As the world expects of British guardsmen, each man has the ramrod bearing born of hours of meticulous parade-ground discipline. **Above right:** Lying to the side of the Gabes-Medinine road in Tunisia, rows of simple wooden crosses mark the graves of the men who died in the 6th Battalion's attack on the Horseshoe. No fewer than 77 guardsmen were killed – if discipline and morale had not held, the Grenadiers would have suffered a far higher toll during their nightmarish ordeal.

carriers. The enemy closed in, taking Wiggin and several other men prisoner. Evelyn was not so fortunate – his body was never found.

At 1000, Lieutenant-Colonel Clive was among the last men to return to the southern side of the Wadi Zess. In the words of one officer, 'I cannot tell you how marvellous he was in the battle, he was quite oblivious to danger. All the men would follow him straight into the jaws of hell if he asked them.' The companies had made it back, and the ferocious battle of the Horseshoe was over. The 6th Battalion had lost 77 men killed, 93 wounded and 109 taken prisoner. The battalion had captured and held its objectives until ordered to withdraw, and had exhibited a courage and determination unsurpassed in the history of the Grenadier Guards.

When fighting finally ended in North Africa, a detachment from the Pioneer Corps led by Sergeant White returned to the Horseshoe to erect a memorial to those who died. Some years later, Colonel Sir Thomas Butler, ex-company commander, arranged for the Mareth Cross to be brought to England. It was a fitting tribute to the gallant men of the 6th Battalion, Grenadier Guards.

THE AUTHOR David Williams is a freelance writer who has contributed a number of articles to military publications, specialising in the history of World War II and the Korean War.

SCOUTING FOR DANGER

THE SELOUS SCOUTS REGIMENT

Men fighting a counter-insurgency war need to be very special, for not only do they have to stalk an elusive enemy, often operating in difficult terrain, but they also have to be self-reliant in the field. They have to be fit, resourceful and capable of working under conditions that push them beyond the limits of normal endurance.

The Selous Scouts Regiment, formed in December 1973, was not the first unit of this type, nor were its members breaking new ground in anti-terrorist methods – small units disguised as the enemy were used by the British in Malaya and Kenya in the 1950s – but the Scouts, in their brief history, became one of the finest exponents of the art.

Their success reflected the quality and quantity of their training. All undercover units undergo strenuous testing but, to many observers who were unfamiliar with the harshness of the Rhodesian bush, it seemed that the Scouts were subjected to almost barbaric trials of strength and stomach. Excessive or not, their training paid dividends in the field.

Inevitably, because of the tight security that surrounded the Scouts' operations, members of the regular forces, already resentful of their 'special' treatment and casual dress, began to question their worth. Things came to a head in the late 1970s with the Scouts being accused of gun-running and poaching. For a time the regiment weathered the storm but, with the resignation of their commander in 1979, it was clear that the end was in sight. In March 1980, following the take-over by African nationalists, the Selous Scouts were disbanded and the unit's short career came to an inglorious end.

A decade of warfare in the Rhodesian bush made the Selous Scouts the most experienced anti-terrorist unit in the world – but even for veteran troopers the war held the promise of certain death if vigilance ever slackened

Left: Scruffy but tough, Selous Scouts, wearing the blue denim fatigues typical of nationalist guerrillas, pose by a bus captured in a cross-border raid into Mozambique. Blackened faces and Soviet-made AK47 assault rifles helped to create the illusion of a rebel force. Below left: Bush skills were a key factor in the Scouts' success. This man is armed with an FN FAL rifle and knife. Below: Heavily armed black scouts on patrol in arid country. The leading man carries a GPMG and his partner an FN FAL rifle.

The Selous Scout Regiment had a short operational history, but under the inspired leadership of Major (now Lieutenant-Colonel) Ron Reid Daly, its members won a fearsome reputation as the best bush soldiers on the African continent. The regiment acted as a combat reconnaissance force; its mission was to infiltrate Rhodesia's tribal population and guerrilla networks, pinpoint rebel groups and relay vital information back to the conventional forces earmarked to carry out the actual attacks. Scouts were trained to operate in small under-cover teams capable of working independently in the bush for weeks on end and to pass themselves off as rebels.

The Scouts were a strictly volunteer force; only highly motivated men of the very highest calibre could fulfil the tasks they had to undertake. A mere 15 per cent of the many that signed up to join the regiment emerged from the tough training programme with the right to wear the brown beret of the Selous Scouts. Reid Daly knew the men he wanted:

'a special force soldier has to be a certain very special type of man. In his profile it is necessary to look for intelligence, fortitude and guts potential, loyalty, dedication, a deep sense of professionalism, maturity – the ideal age being 24 to 32 years – responsibility and self discipline.'

The basic training weeded out the weak and singled out the finest

Each man had to be a loner, capable of living alone in the bush, but also able to work as part of a team. It was essential that basic training weeded out the weak and singled out the finest, most suitable recruits.

Selection for the Scouts was rigorous and even tougher than the SAS course. As soon as volunteers arrived at Wafa Wafa, the Scouts' training camp on the shores of Lake Kariba, they were given a taste of the hardships they would have to endure. On reaching the base, tired and soaked in sweat – the trainers had ordered them to run the final 25kms – they saw no cosy barracks, no welcoming mess tent, but only a few straw huts and the blackened embers of a dying fire. There was no food issued. From this point instructors set out to exhaust, starve and antagonise the recruits. They usually proved so successful that 40 or 50 men out of the original 60 regularly dropped out in the first two days.

Seventeen days of pure hell was the basic course. Every morning, from first light until 0700 hours recruits were put through a strength-sapping fitness programme and barely had time to take a rest before having their basic combat skills sharpened. The day ended with the men having to run a particularly nasty assault course designed to overcome their fear of heights; and then, as soon as darkness fell, night training began.

No rations were issued for the first five days at the camp and recruits had to live off the land. On the third day a dead baboon would be hung up and left to rot in the blazing sun. Two days later it was cut down, gutted and (maggots and all) cooked. Reid Daly explained why:

'Few people are aware that rotten meat is edible if thoroughly boiled – although if reheated a deadly botulism could kill you. Scouts on a reconnaissance mission, where supply might not always be possible, could survive on a rotting carcass, but they had to be made aware of this by practical experience, otherwise they would never have eaten it.'

The last three days of basic training were given over

to an endurance march. Each man had to carry, apart from his weapons and a few rations (125g of meat and 250g of mealie meal), a pack loaded with 30kgs of rock over a distance of 100kms. The rocks were painted green, so they could not be discarded during the march and replaced with others nearer the finish. To make doubly sure of their stamina, the final 12kms was a speed test. This section had to be covered in two-and-a-half hours – it meant the men had to push themselves all the way.

The few men who finished the first stages of training were, after a week's rest, taken to a special camp to undergo the 'dark phase'. If the Scouts were to be effective it was recognised that they would need to look, act and talk like real guerrillas. The base was built and set out as a rebel camp and instructors were on hand to turn recruits into fully-fledged members of the enemy.

'Pseudo-groups' – Rhodesian troops posing as rebels – were the main active units of the Selous Scouts. In the 'dark phase', they were taught to break with habits like shaving, rising at regular times, smoking and drinking, and to adopt a guerrilla lifestyle. Everything from the ritual slaughter of a goat (by partially slitting its throat before strangulation), to walking through the bush in single file, was taught. Vital information on the operational methods of rebel units in the field (such as their preference for arranging meetings by letter), uniforms, weapons and equipment were gathered from killed or captured guerrillas.

Although black soldiers of the Scouts were the

spearhead of any pseudo-group, and had most direct contact with the enemy, white officers tried to pass themselves off as black – at least at a distance. Blacking-up, using burnt cork or theatrical make-up, wearing a large floppy hat and growing a beard, were all devices that helped to hide the more obvious European features.

Scouts were dropped by helicopter and, making use of their bush skills, moved in on foot

The recruits who survived the training programme had little time for self-congratulation. The Rhodesian security forces needed every man they could muster to combat the growing menace of the nationalist guerrillas and hastily formed units known as 'sticks', consisting of one or two white officers and up to 30 black soldiers, were despatched into the bush to seek out the enemy.

In such delicate operations it was essential that the rebels did not know of the Scouts' presence. Unusual movement by road, or air would give the game away, and so Scouts were dropped by covered lorry or helicopter, at night well outside a suspect area, and, making full use of their bush skills, moved in on foot. Once in position, a camouflaged observation post would be established on a convenient hill with a good all-round view. It was at this stage that the Scouts' bush skills came into their own. For the men on the ground there would be no resupply for many weeks; they had to find food, live undetected, track

Left: High-wire act. Assault courses were designed to conquer fear of heights, improve agility and build stamina. Below left: Combat wear was casual. These men, armed with FN FAL rifles, are wearing non-regulation shorts and canvas training shoes. Below: Tense moments as Scouts on a raid take up defensive positions while a punctured tyre is changed. Below centre: A deep penetration team is picked up by a Rhodesian Air Force Bell 205 helicopter. Bottom: Another rebel supply dump goes up in flames after a Scout attack.

and make contact with the enemy without revealing their real identity.

The Scouts used information gathered by attached police Special Branch units to make contact with the guerrillas. While the white officers stayed at the observation post, emerging only at night to hear reports, give orders and relay any valuable information back to base, black Scouts would move into a village disguised as rebels and try to meet the local contact-man. Contact-men gave the guerrillas food, shelter and information. It was often an easy matter to find the right man; the Scouts always had good intelligence on the enemy and several of them were 'turned terrorists' – former rebels who had been captured, made an offer they could not refuse and sensibly decided to join the regiment – who had first-hand knowledge.

The usual drill was for the Scout group to be accepted by the contact-man and, once they were, to arrange meetings with other local guerrilla groups

CHIREDZI: A HUNTER-KILLER OPERATION, 1976

On 18 April, a Selous Scout patrol of three men, led by Sergeant Lucas, became suspicious of an overly noisy gathering of local villagers.

With great coolness the Scouts eased their way through the crowd until they reached the front. The speaker was pumping his AK sub-machine gun up and down above his head.

Lucas knew it was a cue he could not miss. Calmly and unseen he raised his rifle, took sight on the terrorist and fired. The bullet struck the man in the centre of his face.

For a frozen moment there was absolute and total silence – nobody in the audience could begin to believe what happened – then pandemonium broke out and the remaining terrorists broke and fled in every direction.

Sergeant Lucas and his companions, taking advantage of the confusion, slipped back into the cover of the night, radioing a report back to base when they were clear.

at a particular time and place. It was not always that easy, as many contacts were suspicious of their unannounced guests and the Scouts had to go to extreme lengths to prove their loyalty to the rebel cause. On one occasion the Scouts staged a mock night attack on a white farmhouse to convince both the security forces, and the rebels they had contact with, that they were guerrillas. Indeed, they went to the extent of covering the area with blood and dead bodies were provided by play-acting Scouts whose 'corpses', poking out from beneath a blood-stained sheet, convinced everybody of the ferocity of the battle. On another occasion, a white officer pretended to be a prisoner of his black soldiers and endured several beatings to convince the rebel troops of his men's commitment to the nationalist cause.

Meetings were often set up, via the contact-men, with local rebels, meetings which were the opportunity to strike at a band that had infiltrated Rhodesia, but the Scouts themselves never attacked guerrillas if they could help it. Their firepower and combat skills with captured equipment were more than a match for that of the enemy, but all their hard work in getting themselves accepted in a particular area would have been wasted if they took a hand in any action to destroy a guerrilla force. Reid Daly decided that the killing had to be done by helicopter-borne units of the regular army. 'Fireforces', consisting of a helicopter gunship, three troop-carrying helicopters and a paratrooper Dakota, kept the appointments set up by the contact-men. They would arrive suddenly and wipe out the enemy.

Many of the officers and men of the regiment came from the rural areas of Rhodesia and were past masters at tracking, but it was not always an easy matter to switch from hunting wild animals to stalking the far more dangerous terrorist groups. On cross-border reconnaissance, Scouts tracked guerrilla units for anything up to a week, searching for signs of guerrilla activity, especially in the morning or early evening when the sun's slanting rays highlighted even the slightest sign of movement. They paid particular attention to any evidence of disturbed vegetation and kept a careful watch for the sole-prints of the enemies' shoes in the dust.

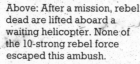

Rhodesian-born Lieutenant-Colonel Ron Reid Daly joined the army in 1951 when he volunteered to fight, with C (Rhodesia) Squadron of the British SAS, against communist rebels in Malaya. After transferring to the Rhodesian Army he worked his way up through the ranks and became a Regimental Sergeant-Major. In 1973, as a captain, Reid Daly was persuaded by General Walls, the chief of the Rhodesian Army, to form a regiment that became known in 1974 as the Selous Scouts – an elite special forces unit that was urgently needed to combat the growing threat posed by nationalist guerrillas. Drawing on his Malayan experiences Reid Daly built up a skilled and highly professional regiment from scratch, but though it performed magnificently in the field, its unorthodox methods won him few friends in the regular army. Reid Daly had several brushes with the military authorities, and in 1979 he was court-martialled for insubordination after being

involved in a blazing public row with his superior, Lieutenant-General John Hickman. A minor reprimand was issued and Reid Daly, no longer able to count on the unqualified support of his fellow officers, resigned. In November 1979 the command of the Selous Scouts was handed over to Lieutenant-Colonel Pat Armstrong and Reid Daly ended his association with the regiment he had worked so hard to create.

Above: After a mission, rebel dead are lifted aboard a waiting helicopter. None of the 10-strong rebel force escaped this ambush.

Any game that was caught or found dead, and plants or roots – Scouts had been taught to distinguish the edible from the poisonous – provided food. Scouts were forbidden to shoot animals as too much noise in the bush might give their position away. Fires, if lit at all, were made from bone-dry kindling that did not give off smoke. At night Scouts dug a 300mm-deep pit in the ground to hide the fire – even the tiny flickering of a dying ember can be spotted at up to 800m in the dark, and this could have the most disastrous consequences.

All hell broke loose and the Scouts opened up with everything they had

By making full use of Reid Daly's ideas the Scouts managed to maintain their cover and severely damage the African nationalists' war effort in Rhodesia. It soon became clear, however, that bases in neighbouring African countries such as Mozambique and Botswana still posed a major threat to state security. Small units of Selous Scouts, working in conjunction with elements of the regular army, were ordered to carry out a series of cross-border raids. Columns of armoured cars, troop carriers and buses penetrated deep into enemy territory.

The most famous raid, carried out with the Scouts' usual skill and daring, was on the rebel base of Pungwe/Nyadzonya in Mozambique. In August 1976, 72 Scouts, in 10 Unimog trucks and three Ferret armoured cars, attacked over 5000 guerrillas. They calmly drove into the camp where they were welcomed by the enemy. Suddenly the rebels realised

their mistake and all hell broke loose. The Scouts opened up with everything they had, and by the end of the battle some 1200 rebels had been killed, while only five Scouts were wounded. Many other raids of this type were carried out before the regiment was disbanded in 1980.

Reid Daly's unorthodox approach to counter-insurgency operations made him extremely unpopular with officers of the regular Rhodesian forces. Several high-ranking officers believed that the Scouts were more trouble than they were worth and had, on occasion, endangered the lives of members of the army. Worse was to follow, on 29 January 1979, all Selous Scout operations were cancelled after a bugging device had been found in Reid Daly's office. Two days later he launched a personal and public attack on the army commander, Lieutenant-General John Hickman, that resulted in a court-martial. Although Reid Daly was given only a minor reprimand, he resigned his command.

In their short history the Scouts inflicted great losses on the enemy yet, because of the inevitable secrecy that surrounded their operations, few Rhodesians knew of their existence. It was not until the end of the war, when Combined Operations, Rhodesia, issued a statement that credited the Scouts with responsibility for 68 percent of all rebels killed, that the scale of their success was publicised. In less than seven years of almost continuous combat the Selous Scouts lost only 36 men killed in action, but had accounted for several thousand guerrillas.

THE AUTHOR Peter Stiff was an officer in the Rhodesian Police Force for 20 years, and has written a number of authoritative books on the Rhodesian War. Two best-sellers, *Selous Scouts Top Secret War* and *Selous Scouts – A Pictorial Account*, are published by Galago Publishing (Pty) Ltd, PO Box 404, Alberton 1450, Republic of South Africa.